IN BUSINESS AND BATTLE

Annabel,

with love from

from Dad

xx oo .

In Business and Battle

Strategic Leadership in the Civilian and Military Spheres

Edited by

Charles Style
Formerly Royal College of Defence Studies, UK

Nicholas Beale
Sciteb, London, UK

David Ellery
Foreign and Commonwealth Office, UK

GOWER

Gower Applied Business Research
Our programme provides leaders, practitioners, scholars and researchers with thought provoking, cutting edge books that combine conceptual insights, interdisciplinary rigour and practical relevance in key areas of business and management.

Published by
Gower Publishing Limited
Wey Court East
Union Road, Farnham
Surrey, GU9 7PT
England

Gower Publishing Company
Suite 420
101 Cherry Street
Burlington, VT 05401-4405
USA

www.gowerpublishing.com

British Library Cataloguing in Publication Data
In business and battle : strategic leadership in the civilian and military spheres.
   1. Leadership. 2. Command of troops. 3. Strategy.
   4. Strategic planning.
   I. Style, Charles. II. Beale, Nicholas. III. Ellery, David.
   158.4-dc23

ISBN: 978-1-4094-3377-4 (pbk)
ISBN: 978-1-4094-3378-1 (ebk)

Library of Congress Cataloging-in-Publication Data
Style, Charles.
   In business and battle : strategic leadership in the civilian and military
   spheres / by Charles Style, Nicholas Beale and David Ellery.
      p. cm.
   Includes bibliographical references and index.
   ISBN 978-1-4094-3377-4 (pbk) -- ISBN 978-1-4094-3378-1 (ebook)   1.
   Leadership. 2. Strategy. 3. Strategic planning. I. Beale,Nicholas. II. Ellery,
   David. III. Title.
   HD57.7.S795 2011
   658.4'092--dc23

2011040100

Printed and bound in Great Britain by the
MPG Books Group, UK

Contents

V

Preface

David Ellery

'*In Business and Battle*' represents the interaction between top strategic leadership in the commercial, civilian and military spheres. It is the fruit of the Royal College of Defence Studies (RCDS); a unique institution with an unrivalled international reputation for excellence in analysis and commentary on global strategic issues, underpinned by the extensive personal experience and expertise of senior personnel from around the world preparing for the highest responsibilities in their respective spheres – all members of its annual programme. The stimulus for this book was the 2009 programme – a top-vintage cohort, representing some fifty countries, and who shared over a thousand years of experience in some of the most complex and demanding environments on earth.

This is a work which harnesses this experience and expertise, it is represented by six syndicates – all comprising outstanding members of the programme – who tackle specific chapter-themes, each of which illuminates critical aspects of contemporary top strategic leadership, and provide a personal commentary on them. Dovetailing with these are chapters by leading practitioners in the commercial field, who made uniquely insightful and highly acclaimed presentations to the programme. My two editorial colleagues, both with long-standing experience of operating at the sharp-end of strategic leadership, supply the Introduction and Conclusion. Together, the individual chapters grip and elaborate a wide spectrum of current and emerging challenges for top strategic leaders and offer personal perspectives on the attributes and approach of such leaders.

The Editors wish to express their enormous gratitude to Michèle Bangham for her indispensable contribution to the book in the form of proof-reading, advice on the text and overall production support.

The views and opinions contained in this book are wholly those of the individual contributors to it: they are not, nor do they seek to be, in any way expressions of governmental or official departmental attitudes or policy.

Foreword

Sir John Parker FREng

Throughout my executive career, at Harland and Wolff, British Shipbuilders and Babcock International, I was involved in an industry that has close links with the nation's defence. As a non-executive Chairman my companies, principally National Grid, RMC and Anglo American, have been civilian but serving on the boards of GKN and now EADS. I have retained an interest in companies with strong links to defence. I was therefore happy to accept the invitation from Nicholas, Charles and David to write a foreword to what I believe is the first book to bring together and contrast the issues of top strategic leadership in the defence and industrial spheres.

People used to talk about the 'military-industrial complex' but for a variety of reasons there is very little movement of people between the two spheres. There are striking differences in culture: certain aspects, such as defence procurement, are extraordinarily complex and very different from their civilian counterparts, and financial controls operate quite differently. People from industry are, understandably, not given high military command, and there is only one FTSE 350 Chairman (Sir Colin Terry at Meggitt) who was previously a senior military officer. Nevertheless, there are many points in common in the issues that strategic leaders face, some fairly obvious and some less so.

One point I would highlight is the multinational and multicultural aspects of top strategic leadership in both the military and industrial spheres. With very few exceptions, major military operations have been multinational coalitions since at least 1914. Leading coalitions in what are now described as complex ecosystems is a fundamental requirement for top management. This is also an increasingly fundamental reality for businesses in the twenty-first century: complex supply chains, which are so intricately linked to a major-project company lead and consortiums or joint ventures that are formed to execute significant defence contracts.

The RCDS, with its unique blend of senior military officers and other officials from all over the world being prepared for top strategic leadership positions, provides a unique vantage point from which to bring together this book. Everyone who is interested in strategic leadership has something to learn from these chapters, and a deeper understanding of the challenges and opportunities facing strategic leaders can only be helpful. As has been demonstrated time and again, whether in companies, armies, schools or other fields, the quality of strategic leadership is vital to success.

One common feature links the success of either industrial or military leadership, and it is the ability to communicate your strategy to those that must execute it. It has been said that communication is the sister of leadership; or to quote the Chinese proverb: 'When the best leader's work is done the people said "*we did it ourselves*".'

PART I
Introduction and Setting the Context

Introduction

Nicholas Beale FRSA
Management Consultant and Social Philosopher

It is said that Montgomery was asked to give a talk on the fundamental principles of strategy, and declined, on the grounds that there was only one fundamental principle of strategy: 'do not march on Moscow'.[1] It was a fascinating contrast to hear Ian Davis' masterly overview of the principles of business strategy which is summarised in Chapter 2. Conceiving and editing this volume gives us a chance to reflect on the fundamental similarities and differences between business and military strategy.

The most fundamental difference is that almost all businesses are 'engaged' every day, at least during business hours: fighting for customers; market share; physical and intellectual assets; and for money and profits. By contrast, actual fighting only occurs for a very small proportion of military time. British Forces have seen more action than most, with some engagement in combat operations during at least 220 of the last 250 years. But with a few notable exceptions these engagements have only occupied a tiny fraction of the military personnel, and even during the Second World War the proportion of troops,

---

1 But in a speech in the House of Lords (30 May 1962) he said: 'Rule 1, on page I of the book of war, is: "Do not march on Moscow." Various people have tried it, Napoleon and Hitler, and it is no good.'

sailors and airmen who were actually in combat at any one time was seldom more than 10 per cent.[2] This leads to another fundamental difference, which is that the strategic landscape changes much more rapidly in a business context than in a military one. The five nations that were Permanent Members of the Security Council when the UN Charter was signed in 1945 remain in 2010 the world's major military powers. By contrast, of the 30 members of the Dow Jones Industrial Average, only five of present constituents were members in 1945[3] and five of the present members did not exist in 1945. Furthermore, the second and third largest companies in the US by market value, Apple and Microsoft, were founded in 1976 and 1975 respectively. Which points out a third fundamental difference: it is much easier to find reasonable metrics for business success than for military might. Although defining and measuring business success is not as remotely a simple or straightforward as simply totting up profits or 'market cap', there are a cluster of reasonably well-understood metrics. In the military sphere, it is not always clear who has won a battle and it is rarely clear who *would* win a war or battle before it is fought – if only because a side that expects to be defeated generally tries to avoid the conflict, especially since Trafalgar.

So what are the main similarities? I think the key common feature is that the strategic leader must achieve results

---

2 This is not of course to suggest that military personnel who are not actually in combat are not playing a vital role. Especially in technically sophisticated modern warfare, the ratio of support to front line personnel can be high. Nor here am I trying to make a point about bureaucracy. Parkinson famously observed that the number of civil servants at the Admiralty increased quite independently of the number of ships, and that commercial organisations have similar dynamics for increasing bureaucracy. Guy Hands, when he bought EMI estimated that for every person involved in their front-line activities there were over 20 in other roles.

3 DuPont, ExxonMobil (as Standard Oil), GE, P&G and United Technologies (as United Aircraft).The members that were founded after 1945 are Cisco, Home Depot, Intel, Microsoft and Walmart. Verizon, which was originally part of AT&T, is now a separate entry.

through long chains, over a large sphere of action (in time and space), with minimal direction. Consider a strategic leader, such as a 4* General/Admiral or the CEO of a large concern, and contrast her or his situation with that of a tactical leader such as the Commander of a battalion/warship or the General Manager of a business unit. They both need vision, determination to get results, communications skills and the ability to inspire. They both need to be able to take decisions and to have sound judgement. However, a Tactical Leader can, and indeed should, know most of or all the people under his or her command, and certainly should know all the managers for whom he or she is responsible. The Tactical Leader's sphere of action is usually pretty limited in time and space, and when it is not (such as the head of the IT department), it is highly constrained by function. By contrast, the Strategic Leader invariably has to achieve results by leading leaders of leaders (or by managing managers of managers), often with more than three levels between him or her and the people at the sharp end. The Strategic Leader will have a wide geographical sphere of action[4] and will also be taking decisions where the time dimension is very considerable. Not only will the decisions taken have repercussions over many years, but the strategic leader's ability to make decisions will be constrained by those taken many years ago by his or her predecessors.

Strategy always works within a complex ecosystem and in large concerns this ecosystem is increasingly global. Almost everybody in the economic world has at least one mobile phone. This will typically contain software designed in the USA, chips designed in the UK but manufactured, like the phone itself, in Asia, and components made from minerals from Africa and quite probably Australia and South America.

---

4 There is an interesting discussion in Peter de la Billiere's 'Storm Command' about how he fought to have his area of command extended outside the Gulf (pp. 55–57).

Many other products have similar global provenance. Even more immediately, movements in the financial markets have tremendous influence on the actions of both governments and companies. Of the ten countries with the largest Foreign Exchange Reserves: six are in Asia, the rest are Russia, Saudi Arabia, Brazil and Switzerland; and with the ten largest Sovereign Wealth Funds three are in Asia, three in the Middle East, and the others are Norway, Russia, Australia and Libya. The days in which economic power was largely in the hands of the USA and the EU are well and truly over. Another aspect of the ecosystem which concerns top strategic leadership is the information ecosystem: most visibly in the media and web-based communications. It is much easier to predict and quantify the scale of an environmental disaster than the scale of the PR disaster that follows it, which clearly can inflict more economic damage to a company. In warfare strategic leaders are also highly constrained by public opinion and news images.

The Tactical Leader works within a framework set by his or her superiors, who can generally be consulted about a difficult matter and will require decisions above a certain size to be referred upwards. Strategic Leaders are never entirely free from such constraints, but it is a matter of degree. In many countries the CEO is not the same person as the Chairman of the Board and CEOs are always subject to Board oversight, at least in theory. Except in a military dictatorship all officers, however senior, are subject to some forms of political control. Nevertheless, the degree of autonomy of a Strategic Leader is much higher than that of the Tactical Leader. This, by the way, is one reason why the wise Strategic Leader always has at least one trusted confidant who can be a sounding board and with whom he or she can have effective and honest strategic discussions. In many companies the head of HR performs this function to a significant extent, since the Strategic Leader has to lead through her or his top team and the incentives and messages they give, so appointing the right people in the right roles and giving them the appropriate incentives is a major aspect of the Strategic Leader's job. Strategic Leaders generally need to be subtle about getting the right mix of constructive feedback from their subordinates, because Strategic Leadership is always a team effort, a point well emphasised by Dr Bertling in Chapter 3.

Some important lessons about strategic change can be drawn from material that was presented to the RCDS Members but which for various reasons are not part of this volume.

The first is that it is extremely difficult to get real strategic change in a large organisation unless there is a serious sense of threat. The great success of many organisations can often be traced to a major setback. Apple, for example, is currently an outstanding success. But in 1997 it was on the verge of bankruptcy and only after Steve Jobs, who had been ousted in 1995, re-joined as CEO did it begin to reinvent itself so spectacularly. Rolls-Royce actually went through bankruptcy proceedings in 1971 and following this traumatic experience became one of the world's most successful aero engine companies, and the only one that is not part of a large conglomerate. It is almost impossible to make radical changes without a real sense of threat that is understood widely enough in the organisation.

The second is that culture can defeat strategy. Organisations often embark on strategic change programmes without really appreciating the extent to which the realities of organisational culture will defeat the neat rationalisations of strategy, even if it is well defined and relatively clear. This is partly because, outside the realm of battle, real success in an organisation is subtle and context-dependent. Even in the world of publicly quoted companies it is never as simple as 'Profits' or 'Share price' or even 'Total Shareholder Return'. These abstractions are generally quite inadequate to motivate people, and in addition it is perfectly possible to boost any one of these measures in ways which are to the detriment of the firm's well-being. People's behaviour is transformed by how they think and how they interact with each other: these are essentially cultural issues. Any strategic change will therefore tend to require a culture change and at least a partial change of personnel amongst the 100 top leaders and many of their direct reports. This has to be done very carefully. It is no good fixing a strategy and then having a 'Culture Change Program' to enforce it. Although it is sometimes interesting to think about what the 'Ideal' strategy might be for an organisation the only successful strategies are those that an organisation can successfully execute. A company run almost entirely by engineers, mathematicians and physicists is unlikely to become an FMCG business no matter how logical it might seem for it to do so.

This brings us to the third point: the need for simultaneous change. Achieving real strategic change requires a remarkable blend of vision and detail and an ability to coordinate changes on many levels simultaneously, or almost simultaneously. Getting the balance right is very tricky. Except in wholly exceptional circumstances, a Year Zero approach won't work. Leaders need to work sufficiently with the grain of their organisation, whilst being ambitious enough to make changes where they are most effective. This requires extensive and clear communication and a message which is rich enough to deal with the reality of the organisation but simple enough to be communicated. Communication needs to be extensive, coherent and regular. It is also important that it should be two-way. The difficulty for strategic leaders in getting honest feedback within their organisations is notorious, and it is extremely important to find reliable channels for this.

Fourthly, we should remember that leadership is context dependent. The qualities needed to push through a major change may not be those that are required to lead an organisation to yet another stage of development. Although over-rapid rotation of leaders is a serious problem, excess longevity at the top also creates major difficulties. If really able people who are capable of top strategic leadership positions are kept waiting too long they will leave, and the organisation will then have grave difficulties in adequate succession planning. Something like seven years appears to be about the right length of tenure for CEOs and although one should not be too rigid about this, anything over 10 years usually becomes pretty problematic. One of the great advantages of separating the role of the Chairman and the CEO, which is commonplace in the UK and becoming increasingly accepted elsewhere, is that changes in the top leadership are less de-stabilising if the Chairman stays in place. CEOs are often very reluctant to let go, but in many respects being CEO is one of the worst jobs in any organisation, and being Chairman is one of the best. So, although it is a wrench, handing over to new leadership is a vital responsibility and a key criterion of success for a strategic leader.

The remaining chapters are written by international syndicates of RCDS Members, reflecting on their extensive experience, expertise and studies, coupled with their engagement with strategic leaders, including their reactions to the talks from Ian Davis and Dr Bertling. Chapter 4 explores the uniqueness of top strategic leadership, and Chapter 5 looks at the characteristics of top strategic leadership. Chapter 6 looks specifically at the multinational and multilateral context, talking illuminatingly about 'Coalitions of the Variously Willing' and quoting Churchill's aphorism that 'there is at least one thing worse than fighting with allies – and that is to fight without them'. Chapter 7 looks at strategic decision making, drawing on some of the behavioural insights that have emerged in the last few decades. Chapter 8 focuses specifically on the vital task of talent development. My fellow-editor Vice-Admiral Style, the Commandant of the RCDS, then sums up in a wide-ranging review with deep personal reflections based on decades of leadership and developing leaders for the future.

A book like this can only scratch the surface of such a vast topic and stimulate thinking. But with 25 contributors from 11 countries spanning a range and diversity of current leadership positions in the private, public and military spheres, we hope it offers at least some valuable strategic insights to both military and industrial leaders, and to those who are intrigued by the complex and vital decisions that strategic leaders across key sectors of society have to take.

Finally, as a civilian perhaps it is easier for me to mention a salient difference that is both obvious and fundamental, but should not go unacknowledged. Military leaders are faced with decisions that put lives on the line, sometimes on a very large scale. In combat, troops are given orders which may well lead to their deaths, and if things go wrong thousands, or tens of thousands, of civilian and military deaths can occur. Very serious physical and mental injuries are also inevitable in war, and sadly the UK has not been as diligent at honouring the Military Covenant as we should have been. Military leaders all know people who have been killed and seriously injured in the line of duty, and almost without exception will have put their own lives on the line somewhere in their career. Equally, because military leaders know the horrors of war, they are generally keen to avoid unnecessary conflict. One of the major benefits of the Royal College of Defence Studies is the close working relationships and informal links that are built up between Members from different countries. Among some of these hostility is traditional and has led to war between them. I don't suppose we will ever know how many conflicts have been avoided by wise use of such informal contacts, but I suspect quite a few. Thus both in combat and the avoidance of combat, the security that we experience and that allows complex and successful international businesses to flourish depends, to a very large extent, on the dedication, professionalism and courage of our servicemen and women and their leaders. We owe them a great deal: this is a debt that should be honoured and never forgotten.

① The Current Strategic Environment

Captain N. Stanley RN
Rt Hon Lord Cope of Berkeley
Ms D. Di Carlo
Major General N. Khokhar

'A smooth sea never made a skilled mariner.' (English proverb)

The strategic leader sinks or swims in a strategic sea. Although few leaders will be entirely isolated from the day-to-day business that affects us as men and women, the capacity to generate or inspire strategic vision, and to take the key decisions that mark a strategic leader, must be applied to – and within – a context that sits well above the humdrum. Put simply, while the strategic leader may well feel the same sense of impotence as non-leaders in the face of the recalcitrant bank clerk or uncooperative petrol pump attendant, he needs to be comfortable with the contextual levels at which national and international financial policies operate, or where geo-politics impact upon the price of petrol at the pumps. Sometimes this level of understanding will have to be of a highly specialised or focused nature linked to the leader's specific field of activity. However, to an increasing extent the strategic leader needs a span of intellectual awareness and insight that is attuned to a contemporary strategic environment that presents novel challenges.

This environment is one of unprecedented complexity. Its nature is marked by high levels of interconnectivity and mutual dependency, which render useless all simplistic 'cause and effect' calculations. This complexity is exacerbated by a proliferation of non-traditional participants. Shifting perceptions of what had been relatively 'fixed' concepts of security, rights, statehood, success and corporate responsibility add to the challenge, together with the impact of modern communications technology and the media. These trends are shadowed by the probability of strategic shocks, such as climate change, global pandemics, economic crises, use of weapons of mass destruction. None of these issues can be entirely ignored, and neither can they be entirely isolated.

GLOBALISATION

Definitions abound, but broadly speaking globalisation represents the evolutionary process of linking markets, communities, societies, cultures, states and regions through developments in transport, communications and the movement of people. In its current form it is typified by its geographical extent – it is now *truly* global – and by the immense span and sophisticated nature of the mutual interdependencies involved. These include hard security concerns, economic power, trade, international finance, international law and human rights.

INTERDEPENDENCIES

The recent economic crisis demonstrates the difficulty of isolating strategic activity. Although variations in economic models filtered or magnified its impact, each of the world's economies has been affected in some manner. This has either been a direct consequence or as a result of second and third order effects on activities, such as trade and international development. Hence the shockwaves from the global economic crisis have affected, in turn, the international finance system, state economic prosperity and viability, national elections, internal state order and security, international trade, development aid and personal charity donations. Following this pattern, some of the poorest communities in the world were left vulnerable despite their being far removed from the source of the crisis in both a geographic and economic sense. Similarly, the nature of the global system also means that the very richest countries are seldom insulated from the consequences of what happens in distant, less affluent regions. Many fragile states in Africa are home to resources that are vital to the continued viability of the developed world[1] and increased human suffering in these underdeveloped regions can result in internal unrest, which can in turn prevent access to key materials and so heighten tensions between developed countries competing for them.

THE GLOBALISED SYSTEM

Trade and economic interdependencies, together with traditional security concerns, have always been drivers for a functional system of international relations. The challenges

---

1 The columbite-tantalite ore 'Coltan' is mined in the Democratic Republic of the Congo, and is essential to the world's mobile phone industry. DRC is home to around 80 per cent of the world's reserves of Coltan.

that the system must now address have been amplified by accelerations in trade and economic interaction, and fears over the proliferation of nuclear capabilities and other means of mass destruction. Although the highly-polarised client state structure that emerged during the Cold War has lapsed, hopes for a fresh international consensus and the renewed empowerment of the United Nations have not survived contact with the twenty-first century. Fears of a clash of civilisations have not been borne out, but regional groupings are asserting themselves to a greater extent,[2] security alliances continue to wax and wane, and the progressive emergence of a multi-polar world means that newer powers rightly expect to have their voices heard and their views heeded. For good or ill, this cannot help but complicate international discourse, international trade and the search for international peace and security. In the Occident at least, where fears of state on state violence in the short term have eased, the very concepts of peace and security are also undergoing subtle adjustment to reflect recognition of the heightened importance of economic and internal security and their contribution to national well-being and human security. This important shift, which introduces levels of nuance beyond those associated with traditional defence/security models, has been readily apparent in a number of security strategies published over the past five years.

---

2 Examples include the Organization of American States' solidarity over recent UK offshore activity in the South Atlantic, the growing influence of the EU's 'soft power', and the African Union's efforts to maintain primacy in security matters affecting the African continent.

ACTORS

INTERNATIONAL AND SUPRA-NATIONAL

Within the international system, the UN – despite its limitations – exerts a strong moral influence and has occasionally shown its active preparedness to override the needs or wishes of an individual state. Despite their individual limitations, regional defence and security organisations can also facilitate or place curbs on a country's freedom of manoeuvre. Meanwhile, supra-national bodies of a global or regional nature, such as the World Trade Organization and International Monetary Fund, created to support economic, finance and trade development, have become increasingly influential; reflecting a heightened focus upon economic well-being, while helping to oil and regulate the machinery of globalisation. International legislative bodies, such as the International Criminal Court and the International Court of Justice are flexing their muscles, while the sheer economic heft of the major multinational companies – exceeding that of many nation states – has given them an influence over governmental decision making that has brought them firmly into the strategic arena.

STATE AND SUB-STATE

The diaspora of communities, cultures, ideas and creeds that has been facilitated by modern transport and communications has also resulted in the embedding of distinct, separate value systems within host states. Consequently, countries find it increasingly difficult to speak with one voice since primary allegiances within can often be paid to distant figures without. For many Tibetans, the Dalai Lama is a focus for religious and cultural leadership; while closer to home in the

UK, some groups may well see their principal loyalties resting with religious or cultural figures in Africa, the Middle East or the Americas. Through overlooking such significant societal factors, which run from the macro to micro level, a leader runs the risk of dangerous miscalculations if a simplistic approach is taken to strategic decision making.

The 'coal face' of strategic business has also witnessed the emergence of strategic actors whose influence has grown exponentially over recent years. Within the democratic system, lobby groups have steadily grown in their ability to persuade, pressure and influence. The economic crisis will have sharpened their appetite, while climate change has offered environmental lobby groups a secure foundation for continued forays on related issues. In the security environment there has been a proliferation of non-governmental organisations who have assumed or been granted portfolios of responsibility for the alleviation of humanitarian suffering, and a sharp increase in the numbers of private military and security companies. For the military or other government departments, aside from the tactical or operational complexities that the presence of such organisations can introduce to the battlefield, they also have the potential to exert strategic levels of influence through the auspices and synergies of two of the most recently emergent strategic actors: public opinion and the media.

MEDIA

Although the pervasive nature of the media and modern communications is such that few societies are entirely insulated from their impact, the levels of access to these facilities in the Western democracies, together with the nature of the prevalent political systems, has rendered strategic leaders particularly exposed to their direct and indirect influence.

While the mobilisation of the public by the media is not a novel development,[3] the diversity of means, the volume and immediacy of media traffic, combined with a public which recognises itself increasingly as a legitimate stakeholder in high-level affairs, are such that consultation and transparency are increasingly demanded. Away from the more traditional forms of media, modern social networking sites allow any individual to publish their views on 'leadership activity' to a global audience transcending national or cultural boundaries, thus allowing 'everyman' to achieve strategic influence in areas from which *he* previously was excluded through absence of knowledge and absence of voice.

INFORMATION

Underpinning the waxing influence of the media, the public and the individual, are the extraordinary levels of data, data exchange and data access which constitute another key feature of the current strategic environment. Although such developments have tended to be evolutionary, computer and microchip technology have resulted in an unprecedented surge in the amount of processed and unprocessed information that is available to the strategic leader. This acceleration in access, combined with the nature and extent of the material, means that human or technological systems for the management of information are frequently overwhelmed. The challenges that this issue poses will be dealt with later in this chapter.

---

3 William Russell's reports for *The Times* during the Crimean War (1853–1856) caused significant concerns for the British Government.

CURRENT AND EMERGING ISSUES AND CHALLENGES

A non-exhaustive list of specific challenges to strategic leaders would take up pages. The thrust of this section will be to examine the key trends that pose problems for the strategic leader, many of which are derived from the particular nature of the strategic environment outlined above.

THE NATURE OF THE STATE

Whereas only a small proportion of strategic leaders will be actively involved in determining the future of the state and international system, all should have an interest in the implications for their own, specific areas of strategic interest. Key challenges affecting the state arise from the nature of globalisation and its implications for concepts of sovereignty and security, from changes in the relative power and influence of individual states and regional groupings within the international system, and from the change in individuals' perceptions of their duties to the state – and vice versa.

Left- and right-hand markers for our globalised future are a world in which there are seas of prosperity and harmony interspersed with the inevitable islands and archipelagoes of disorder – or the opposite. These models pose very different challenges to the nature and role of the state and thus to the strategic leader – whether his subject area is economic, foreign policy, business and trade, defence or ideology. Some trends are rather more predictable though, and the continued emergence of China and India to global power status is set to continue. The likely impacts of this include a return to multi-polarity – or non-polarity – and a shift of emphasis within global trade and economic institutions that have previously followed what is perceived to be a 'Western' agenda. The implications here will be felt by the entire finance and business sector, and the relative importance of the state entity in these spheres will be of particular interest and importance to them. Regional trade patterns are also likely to shift – Chinese engagement on the African continent is already evidence of this – and the UN's commitment to the cause of international peace and security is likely to be manifest in less expansive interpretations of the relevant UN Charter articles.[4]

Internal changes to the nature of the state are also in evidence. Empowered by information technology and aware to an ever-increasing extent of being part of an international community, individuals have redefined their relationship with *their parent state* and expect or demand its actions to be aligned with their closest individual and communal needs, be they economic security, physical security or value-based. This expectation is filtering through all societal strata – at various speeds and levels of intensity across different countries. Consequently, the individuals who work within state organs, corporations and businesses tend to see themselves increasingly as stakeholders – with a right to a view and a right of expression. Modern communications and communications technology can also result in greater pressure – internal and external – for states to act or take a stance and for these actions and the decision making that surrounds them to be more transparent. As will be discussed later, the new cycle and the perception of a need to demonstrate swift success can also weigh heavily upon strategic calculations.

As a result of the external and internal changes outlined above, the state – the foundation for strategic dealings on

---

4   Including further questioning of concepts, such as 'the right to protect'.

the international market – is continuing to evolve and the manner in which it evolves, both generically and specifically, will have profound implications for all strategic leaders.

FRICTION

Although the future nature and role of the state is probably the most weighty background issue facing strategic leadership, the more immediate challenge is likely to be that of 'friction'. The term was used by Clausewitz to capture the myriad of factors that stood between the articulation of a plan and its successful execution, and consequently the difference between theoretical and 'real' war. The concept is valid for any form of strategic undertaking in the contemporary environment where the proliferation of interdependencies and actors is so evident and where single events of relatively minor consequence can find their effects magnified through third- and fourth-order impacts. Some militaries have acknowledged this through the idea of the 'strategic corporal' – an individual whose tactical (local) actions can have strategic implications. Although such individuals may be at their most numerous in a sensitive security environment, they are present in any strategic enterprise and have the real potential to inadvertently derail or dislocate high-level activity. Meanwhile the plethora of related interests that arise from the profusion of interdependencies is such that the charting of cause and effect and, consequently, the identification of a successful strategy or course of action that can surmount the potential for friction is often obscured.[5]

---

5 'Between the idea and the reality, Between the motion and the act, Falls the Shadow.' TS Eliot, *The Hollow Men.*

INFORMATION MANAGEMENT, COMMUNICATION AND THE MEDIA

The technology that has granted us immediacy of communication, access to near-infinite information resources and a 24/7 media culture has presented the strategic leader with crucial challenges. The challenge is manifest in three interrelated aspects: the handling of quantity; the application of a form of quality control; and the identification of the relevant. Each aspect needs to be satisfied if decision taking and direction is to be relevant and informed, otherwise the leader is overwhelmed with data to which no intellectual rigour or sensible prioritisation has been applied – and if everything is a priority, then nothing is.

One of the secondary challenges posed by modern communications arises from excessive reliance upon them as a means of leadership. Efficient? – yes, but only up to a point, and that point is passed where the immediacy of an email fails to convey the nuance and intent that can be achieved by eye or voice contact; and where it fails (in most instances) to impart the vision, direction and drive that are intrinsic to effective leadership and which set it apart from managerial responsibilities.

Finally, the speed and diversity of communications has resulted in a 24/7 media culture in which the pursuit of new 'news' and unique perspectives puts strategic timescales, which tend to be of relatively long duration, under threat. This creates numerous challenges; the desire for transparency is entirely appropriate but can occasionally run counter to strategic enterprises where opacity may be essential. Where public support is needed, media pressure for fresh news over a time span where progress may be relatively slow could stretch tolerance and patience through the absence of a ready, rapid connection between cause, action and effect.

As a footnote to the above, it has to be stated that the innovations in communications have not been entirely to the detriment of the strategist. The need to exploit these developments is touched on later in this book.

SECURITY CONCEPTS

Closely linked to the process and nature of globalisation, the shifting role of the state and the change in individual perceptions, traditional notions surrounding the concept of security are now being challenged. In many countries where traditional, external threats to territory have eased, where freedom of speech is taken as a given and where individual prosperity and health is seen as a right, hard security concerns are being eclipsed by concerns over economic and individual security. While attempts to separate territorial security from economic and human security have never been entirely valid, the degree of convergence is now such that more spheres of strategic activity are taking on explicit and more immediate importance at societal and individual levels. Scattered over a sliding scale of traditional/non-traditional security concerns, these can include: energy security; water security; food security; religious and secular freedom of expression; freedom of movement; security from state observation; freedom to own property; security from terrorism; security from crime; the freedom to bear arms; and the freedom to bear children. Although some of these concepts and issues have already found expression in state-published security strategies,[6] such publications rarely capture the full extent of societal concerns.

---

6   Over recent years, national security strategies for the UK, France and the USA have all broadened their interpretation of 'security'.

Consequently the strategic leader will be operating in an environment where, although the stakes may be higher or lower than those associated with traditional security concepts, their range, nuance, interplay and visibility, combined with greater individual and group interest, will render them more complex than those encountered previously.

SUCCESS

'Zero-Sum' – the theory that gain for one individual or side must be matched by a loss for another individual or side – has long been an underlying concept in competitive relations between states, armies and boardrooms. Assuming a globalised world that lies closer to the more benign of the two models outlined earlier in the chapter, where interdependencies abound and where the paths between cause and effect are seldom straightforward, zero-sum mentality is a danger in anything but the most immediate, existential circumstances. However, although there is an increasing call for value-based or win/win stances, the 'zero-sum' mindset frequently dominates strategic goal setting and strategic decision making. Two extreme examples from the sphere of international relations point to the pitfalls of zero-sum and the benefits of 'win/win'.[7] The post-1919 peace settlement and the failure of Germany to reconcile itself to the aftermath of the First World War is a telling case, but the converse was apparent in the treatment of Japan in the aftermath of the Second World War. Consequently, in most strategic dealings there is a need for a common vision and a preparedness to reconcile and agree upon a common future.

---

7 Best explained in *Non-zero*, Robert Wright.

LEADERSHIP

Underpinning this overview of the strategic environment is the increasingly formidable challenge of leadership. Faced with this environment, its trends and challenges as outlined above, the strategic leader must possess and consistently demonstrate key attributes if they are to succeed at this level – some of these will be innate, while some can be learned and developed. For us these encompass:

- *Vision* – the ability to understand the essence of the strategic environment and to have a clear vision concerning what needs to be done.

- *Breadth* – the ability to think in terms of strategic timescales, resources and key interdependencies in the implementation of the vision.

- *Shaping* – the ability to harness and exploit external and internal actors through promoting a common sense of purpose.

- *Presence* – the ability to make a personal difference in adversity.

- *Communication* – the ability to master relevant information and to engage others through the most appropriate means at an organisational or individual level.

Treatment of such attributes will act as a common thread throughout this book.

PART II
The View from Industry

② Perspectives on Leadership and Strategy

Ian Davis
Senior Partner and former Worldwide MD,
McKinsey & Co. Inc.

I should like to offer my personal perspectives on strategy and leadership.

I want to begin by spreading the canvas, as it were, with a simple and direct statement:

> *'Leadership is contextual; it is multifaceted and it takes different forms in different geographic areas.'*

LEADERSHIP CHARACTERISTICS

I shall set out at this point what I regard as a set of leadership characteristics, which have broad relevance; by this I mean that they are not bound by the particular role or function of the organisation in which the leader operates. For me, the successful leader:

- Initiates action: it is the quality of dynamism that can make them seem 'quirky' in the eyes of others and 'difficult' in their behaviour.

- Sustains momentum: displays resilience and is a 'sticker' – sometimes to a fault.

- Motivates people: I want to emphasise here that this is not the same as 'inspires', since the instrument of motivation might be fear, pure and simple.

- Leads by example: has high visibility within the organisation; the behaviour and attitudes of the leader are closely scrutinised by those they engage with, and who take their cue from them.

- *Shapes* the context, rather than merely *responds* to it: this is the keystone for all the other, preceding characteristics I have touched upon.

STRATEGIC LEADERSHIP CHARACTERISTICS

Let us now take a look at those characteristics which mark out the successful *strategic leader*. These supplement and build upon those characteristics I have just discussed in the context of the *leader* who functions at the lower, more *tactical* level. In my view, the successful strategic leader displays the following:

- Has a clear sense of the timeframe relevant to strategic decision making; this timeframe is always significantly longer than that at the tactical level and is usually closely aligned to the delivery of strategic intent.

- Matches effectively resources and capacity to strategic intent: I want to make it crystal clear that *vision* is not enough; practical application is crucial, since the boundary between vision and illusion can be very narrow.

- Has 'what *can* I do?' as the starting point; musing about 'what *might* or *ought* I do?' is quite simply vapid. Practical, focused dynamism is the watchword.

- Has the ability to spot top talent and to nurture it; this includes the appointment and/or promotion of the right people into important roles *at the right time.*

So, by way of pulling these threads together: these qualities and attributes illuminate a coherence in the attitude and approach of the top strategic leader, whereby the 'why, what and the how?' become inextricably linked.

THE KEY CHALLENGES FOR STRATEGIC LEADERS

The challenges confronting the strategic leader are, of course, many and varied; but for me, the following go straight to the heart of the matter:

- The recalibration of the timeframes with and within which they work.

- The movement from *direct* to *non-direct* leadership. Let us unpack this a little. This includes, crucially, motivating people you don't know: an option here is 'working the network' – or put another way, *influencing the influencers.*

- The delivery of consistently high-quality analysis and planning. These provide critically for the shaping and transmission of a clear message, which people can readily understand and then support.

- The handling of ambiguity: the need to be able to grasp with and manage the 'what if...?'

- The maintenance of a sharp sense of strategic intent and the goals that underpin its delivery. In the sphere of capacity and equipment acquisition, for example, there is the significant risk that for the tactical leader, the acquisition itself becomes the strategic goal, rather than as a means for the achievement of it; to offer an example, this might be the globalisation of the business or the product.

'GROUND TRUTHS'

I choose this heading to introduce some personal insights that I offered to Members of The Royal College of Defence Studies in response to specific points raised by them as part of my presentation to the College in 2009.

THE QUESTION OF HUMANITY AND MORALITY IN STRATEGIC LEADERSHIP

There is no fundamental conflict between them: sustained success in the commercial world is not possible without humanity and morality being core elements of an organisation's strategy. A crucial attendant challenge is keeping the approach to humanity and morality in kilter with the society in which the organisation operates; this helps secure legitimacy, which in turn is a pillar of authority. This is, therefore, an ongoing process – one demanding continual monitoring of relevant developments and shifts within society and acute sensitivity (and adjustment) to them.

TO LEAD OR NOT TO LEAD?

It is not always in the interests of the organisation for its strategic leader to take the lead on every issue. There should be more discrimination and subtlety in play. To *follow* is not necessarily an admission of failure or inability, rather a strategic leader can enhance their authority by choosing judiciously when to let others take the lead and then to support their endeavours, with the aim of better delivering strategic goals. The net effect of this, in other words, is a 'double win'.

THE IMPORTANCE OF LISTENING AND HONEST FEEDBACK

Linked to the above is the matter of listening in an open-minded way to others; being receptive to their ideas and feedback. I wish my message here to be crisp and clear: there should be no mistake or confusion that listening to the counsel and opinions of others necessarily means a demonstration of indecisiveness or the abnegation of leadership responsibilities; far from it, taking cross-bearings and eliciting honest feedback should be routine for the strategic leader – both to test and to hone thinking and approach.

This is why I am an enthusiastic champion of 360-degree appraisal and feedback. It is important not to have too narrow a base for such feedback; it should as far as possible encompass a fully representative range of people with whom the leader operates. And there is much value in supplementing and following-up this feedback by visiting/engaging one-to-one with colleagues and clients, as the providers of it, to obtain first-hand their views and the elaboration of them. Connected to this, I am also a strong proponent of securing a good confidant or mentor – this can be invaluable.

EFFECTING SIGNIFICANT CHANGE - POLITICIANS AS A MEANS TO AN END

There has been an assertion doing the rounds that within government only politicians can effect significant change: I dismiss this as a myth! Civil servants can lead and deliver change by using politicians as tactical allies or, perhaps surprisingly at first thought, even impedimenta in the interests of achieving strategic goals. The challenge for the civil servant is plain: the need is to seize the initiative and to be resourceful in taking it forward.

THE POWER OF PEER PRESSURE

The best organisations are driven by peer pressure, rather than downward or imposed pressure. Through the deft handling of peer pressure it is possible to work on an organisation's culture to enhance its overall governance. This is reinforced by the empowerment of its members, who feel able to comment on and influence its governance issues, and who consequently generate commitment to the organisation and its goals through a sense of ownership of them. In parallel, such empowerment generates dynamism and vitality across the organisation.

So by way of conclusion: from my commercial standpoint, there is no one prescription for successful leadership or the devising of strategy, no magic formula awaiting discovery or application. I can, therefore, only trust that the personal comments and observations I have set out might strike a chord and have resonance in informing the approach of the individual to leadership and strategy.

(3) Commercial Top Strategic Leadership: A Helicopter View

Dr Lütz Bertling
President and CEO of EADS, Eurocopter Division and
member of the Executive Committee of EADS

I have set out in the next few pages my views on top strategic leadership. These are a personal illumination of those things, which I hold to be of significance and that have featured notably in my own career in the upper echelons of the commercial sector.

THE GLOBAL CONTEXT

But before I embark upon this, I should like to place my comments in the current strategic global context.

The end of the Cold War some twenty years ago was the catalyst for fundamental change, the extent and ramifications of which encompass key aspects of national and international activity. I offer three resultant shifts to exemplify this transformed global environment.

First, national interests in the sphere of strategic-level business have all but disappeared. This is a reflection of a global landscape increasingly characterised by issues whose interconnectedness demands a response that transcends the merely national: a coherent international response is the overarching imperative.

Second, in the military sphere, my own experience clearly points up that high-pace development and the speedy delivery of capability are of the essence: the demand for its use is more immediate than ever before.

Third, underpinning this move towards more dynamic delivery of capability is the shift away from large, static-footprint military operations to more expeditionary, Special Forces-like operations. And by way of underscoring the stark character of this telescoping of timescales, it is perhaps enlightening to consider this as analogous to the horse-to-tank transition witnessed in the early part of the twentieth century – a transition which took some thirty years to complete.

THE MAKE-UP OF A STRATEGIC LEADER

Against this global backcloth there is for me a core set of characteristics associated with successful top leadership in the commercial world.

The first of these is passion: the need *to be* passionate and *to be seen* to be passionate about what you do and aspire to as a strategic leader. My own experience unequivocally indicates that although pure intellectual acumen is a key attribute of such a leader, this of itself is not enough: the possession of a high level of emotional intelligence is equally important. These attributes should dovetail neatly and in doing so, critically reinforce each other.

Then there is commitment: the *demonstration* of sustained championing of what the organisation is all about – its strategic vision and the attendant roadmap for the delivery of it. Make no mistake, the onus is on those at the top of the organisation to go the extra mile – to be prepared to give that bit (at least) more than what they ask of those whom they lead. It is no surprise, then, that high reserves of physical, emotional and intellectual stamina are the order of the day – the demands are all embracing.

The characteristics identified above should be underpinned by top-drawer communications skills. These are skills which provide for clear and frequent engagement with relevant parties – crafted and delivered in such a way as both to convey the essential message(s) and to resonate with the recipients. The call here is for a toolbox of communications methods that may comprise, for example, traditional formal correspondence; IT-based media; and direct, personal engagement through 'walking-the-talk'. Visibility and accessibility are core requirements: they bolster the strategic leader's authority.

IMPERATIVES

COMPREHENSIVE TRANSFORMATION

That a strategic leader should lead from the front is axiomatic. If we scratch away at this a little, we discern that the quality that marks out the *strategic* leader is the ability to carry the flag of transformation: to translate the vision of an organisation into a programme of transformation – one that goes beyond structures and processes, and reaches the heart of the organisation and encompasses its culture. Furthermore, the successful strategic leader will ensure that this programme of transformation is not only implemented, but that it is also firmly embedded. This is not possible without deep strength of character, resting on the twin pillars of conviction and courage – the courage to hold fast to what is right for the organisation, rather than the pursuit of narrow self- or factional-interest; and to do so with honesty and humility. Such an approach does more than avoid the bear trap of hubris (and how often have we seen leading lights, not least in the commercial sector, plunge headlong into this trap!). Rather it promotes a preparedness to bring into the open awkward information or data, and to be seen to be addressing it full square. The demonstration of such willpower and the championing of transparency serve to generate trust and inspiration on the part of those whose role it is to offer followership.

TEAM BUILDING AND TEAMS

In today's commercial world, top leaders are top team builders. I am well aware that some will argue, of course, that this has always been the case, albeit to a greater or lesser degree. My contention is that the complexity and fast pace of this world today make effective team building indispensable for success.

The ability to turn business drivers into an appropriate team make-up; to select individuals who satisfy the criteria for team membership, and to nurture and develop them to the point where they individually and collectively perform consistently to the standards demanded of the business without undue direction is at an ever increasing premium.

My experience is that the most effective teams are those whose composition is strong and diverse. Their members readily and consistently apply a critical awareness to the issues they are called upon to tackle; to their associated role and responsibility, and to the overarching vision of the organisation and its wider business agenda. Put another way, positive challenge is at the heart of a successful team. Such challenge is far from a threat for the true strategic leader, rather it is a crucial instrument of successful planning and delivery.

THE DELIVERY OF RESULTS

I am clear that the teams top leaders build and lead, and are there for one purpose and one purpose only: the delivery of worthwhile results. While such results are a sharp business imperative – the achievement of the desired bottom line – the results-oriented approach is also the best way of generating significant and *lasting* satisfaction on the part of team members and the wider workforce; and as such, it acts as a source of motivation and inspiration.

PERFORMANCE OBJECTIVES

Discussion of a results-oriented approach leads naturally to consideration of performance objectives – both individual ones and those of the team. In my view, the strategic leader has the responsibility to set three categories of performance objectives for team leaders. The first should relate to tasks undertaken as an individual; by this I have in mind a contribution that makes the most of personal strengths and qualities relevant to the tasks in hand. The second should be linked to the individual's role as a team leader. And the third category is one as a contributor to the achievement of the performance objectives of colleagues, that is to say fellow team members/team leaders. I regard this latter category of performance objectives as being of high importance: it should account for some 20 per cent of the individual's time and effort. It is without question a crucial component of successful team working, functioning as it does as an inherent counterweight to a narrow 'stove-piped' approach to the business. This by no means precludes 'competition' between teams and team members, but it does act as a break on it becoming unhealthy, to the point where it descends into unproductive (at best) rivalry.

There are many occasions when a strategic leader takes up a post and inherits a team *in situ*. In such circumstances, they should start from a position of trust with respect to the performance of the team: as a rule-of-thumb, I recommend allowing it some six to twelve months to show its paces. The precise timing is, of course, a matter of judgement, which needs to take account of the full range of relevant factors. At the end of this period of 'road-testing' any deficiencies should be gripped – gripped firmly and immediately.

As for the approach that might be adopted by a new member of the team, including the team leader, the thrust here should be in the first instance to build up knowledge of, and competence in the business functions ascribed to the individual personally. Once a solid foundation in this regard has been laid, then the emphasis of effort may profitably shift to the development of effective engagement with the (rest of) the team. After all, professional credibility is at the heart of an individual's successful performance within a team and as its head.

An important aspect of the team function of the strategic leader is that of personal time management. I am developing a software program to identify and monitor a spectrum of issues that appears on my personal radar screen to help determine their importance and urgency and, hence, how I might react to them. It is vital that the list of such issues should not become too long; the point being that those issues requiring personal attention get it and with due timeliness. An important attendant consideration is, of course, what to do in response to those issues that don't qualify for personal attention. An obvious option is to delegate authority for handling them to someone else; this may be appropriate in some instances, but not in every instance: there may be times when it is advisable simply to leave the matter as it lies, if personal engagement is not possible.

THE CALIBRATION OF STRATEGY

So much for team building, team function and team delivery within the context of a set strategy, and the associated role of the strategic leader. But what about the calibration of strategy? The onus firmly rests with the strategic leader to review continually the relevance and effectiveness of the current strategy and to make changes to it when necessary – *but only when necessary*. To put it crisply: the trigger for change at this level has to be both significant and of lasting application. The essential point – which I cannot underscore too firmly – is that the strategy should not become a 'moving target'. As a guideline, a company might conduct a fundamental review of its strategy every three to five years.

AND THE TIME TO GO...

What's my view when it comes to the length of tenure of a strategic leader in any one position? It is simple and consistent. The 'norm' in industry is around four years (and I do not support the current trend to shorten that). In most cases this avoids, on the one hand, an individual staying on too long – past the point where their presence no longer serves the best interests of the organisation concerned and, on the other hand, their being in a post too briefly, which brings with it the risk (not uncommon, I fear) of what I might describe as 'power-point leadership' coming into play. This is where the leader can avoid responsibility for implementing and embedding the 'great' ideas they have proposed and championed upon taking up the post, if not as a means of securing the post in the first place! When all is said and done, it is the delivery of significant and lasting value that counts.

PART III

The Military
Perspective

The Uniqueness of Top Strategic Leadership

Group Captain R. Knighton RAF
Dr S. Akulov
Major General K. Engelbreksten

INTRODUCTION

Every year a large number of books are published on leadership and strategy. Many of them seek to put their finger on what it takes to make strategy or make a strategic leader. But what about leadership at the strategic level? Is it just tactical, day-to-day leadership on steroids or is it a special case requiring special skills? Our view and our experience lead us to believe that it is special and, as we see in other chapters in this book, it requires special characteristics that are not necessarily honed or developed at tactical and operational levels.

WHAT IS THE STRATEGIC LEVEL?

So what is the strategic level? The word strategy is now widely used in all spheres of management and business. The word originally comes from the Greek *strategos*, meaning generalship. The word is formed from *stratos*, which means that which is spread out and *agos*, which means leader. The origin and its use, up until the late twentieth century, were most readily associated with war. The philosopher and Prussian General, Carl von Clausewitz, brought its use to prominence in military terms in his book '*On War*'. For Clausewitz, strategy is the employment of the battle to gain the ends of the war. Clausewitz's writings and thoughts on strategy have been explored by many interested parties and their comments fill many volumes, but in simple terms, Clausewitz saw strategy as the overall plan for war.

In the latter part of the twentieth century, business leaders and writers began increasingly to borrow the term. Since then, there has been an explosion of literature on strategy in business. Universal definitions are hard to find. The term 'strategy' is now used (inappropriately in our view) in myriad circumstances. Even primary aged school children are taught 'strategies' for tackling simple mental arithmetic problems.

To understand the challenge of strategic leadership, we must understand what we mean by strategy and the strategic level. Our rather oversimplified Clausewitzian view of strategy, as the overall plan to deliver the desired ends, is a good starting point. It is easy to see how from here the use of the word strategy has evolved from its original, rather narrow definition. It is now used as a rather grand word for the plan to achieve the objective. Similarly, things are often described as being of 'strategic importance', simply because it affects the next level up, or potentially disrupts an organisation's plan, irrespective of the size of the organisation.

To illustrate the point about the misuse of the term 'strategy' and to explain the issues we are interested in, consider for a moment the owner of a franchised petrol station. The owner may well have a strategy for his business; he may even have a marketing and a financing strategy. But what this book is interested in is the company that owns the brand. This company will be an international or multinational business with a highly complex organisation, employing many thousands of people, with a very long-term perspective, facing considerable risk and uncertainty.

For us the strategic level is the highest level of government or the highest level within large multinational organisations, or those with similar global reach and impact. The strategic level is by its nature and by any measure extensive and long term. And in the modern globalised world, it is also multinational and multi-sectoral.

STRATEGY, STRATEGIC PLANNING AND STRATEGIC LEADERSHIP

The profligate use of the word strategy makes it worth exploring the differences between strategy, strategic planning and strategic leadership. The terms are sometimes used interchangeably and provide a somewhat confusing background for understanding the particular challenges of strategic leadership. In our view, strategic planning goes some way to helping leaders make strategy, but the problem we have found with strategic planning and the plethora of literature on the subject, is the attempt to turn the development of strategy into a science. The tools are helpful but, as we will see later in the book, strategic leadership and the development of good strategy require more than an ability to follow a recipe.

Strategic planning is seen as a process by which strategy emerges. The tendency to try to turn it into a science often leads to the process being described in a rather simplistic and linear manner. Although the process can help guide thinking, truly strategic problems are unbounded, complex and uncertain. The very act of developing a strategy can affect the environment from which the strategy emerged. The level of strategic leadership and the problems we are interested in exist at the boundary between the environment and the organisation or system we are trying to lead.

Strategic planning can be a useful tool in helping us develop a strategy and understanding the tool sets and ability to make strategy are important functions that an effective strategic leader must master, but they are not the only things that a strategic leader must do well. The strategy that the leader develops might set out the plan, but executing the plan, reacting to the shifting context and leading people to deliver the strategy are also key functions of the strategic leader. The strategic leader must be able to understand the environment, determine the direction, develop the strategy and actually deliver the strategy. And delivering the strategy cannot be done without leading the people inside the organisation and influencing those outside it. So, strategic leadership is much more than writing strategy or strategic planning. Strategic leadership is about leading people and organisations. What makes it difficult are the particular characteristics of the strategic level.

THE CHALLENGES OF LEADERSHIP AT THE STRATEGIC LEVEL

Perhaps the most obvious of the challenges the strategic leader faces is one of complexity. The very nature of the truly strategic level means that the organisations that strategic leaders must deal with are extensive. This means in those instances where this term does not translate into a big workforce, it nonetheless implies an organisation which, for example, deals with very long timescales, large geographic reach and long chains of collaboration. That said, while the structure that is established to manage the organisation might be simple, or the business model easy to comprehend, size, particularly where people are concerned, brings with it complexity that stretches the cognitive capacity of leaders. We know from personal experience how communication can be unclear or difficult even in small organisations or unit-like families. As these units are scaled up, the complexity and ambiguity grows with it. Research suggests that we as individuals can only really deal comfortably with group sizes of about one hundred and fifty. Beyond that size we cannot comprehend the web of relationships that a larger organisation develops because we lose the capacity to manage all the relationships on a personal basis. The complexity of the real world, particularly at the strategic level is such that it is impossible for us to know and understand all potentially relevant information. So when we make decisions and lead at the strategic level, we have to manage with imperfect and incomplete information that introduces further ambiguity. For the strategic leader, dealing with big and long-term problems, the uncertainty and ambiguity they face are enormous – the extent of this is a reflection of the nature and scale of the challenges in play.

The truly strategic level, particularly as we have defined it, is also almost certain to include a multinational or multicultural dimension. This further stretches the strategic leader's ability to comprehend the environment in which he has to lead. History is littered with examples of strategic leaders failing to interpret the actions of an opponent or ally because of cultural differences – the US experience in Vietnam springs to mind. We have a strong tendency for cultural mirroring, in which we interpret the behaviour of others using the framework of our own cultural norms and practices. To be effective, strategic leaders must be able to understand the world from more than just their own perspectives if they are to interpret it effectively.

For those involved at the military strategic level, the multinational, multicultural and multi-sectoral dimensions of the problems we face and the solutions that are required to address them effectively are highly apparent. Since the end of the Cold War, militaries around the world have faced complex conflicts that have required the widest possible understanding of ourselves, our allies and our enemies. Afghanistan is the most recent example, but there are many others. In terms of the multinational dimension, coalition forces engaged in operations in Afghanistan have a complex set of national constraints that determine the contribution they can make to the campaign. The need for different nations to work together effectively runs almost from the lowest tactical formation up to national parliaments. The multicultural perspective manifests itself on the ground. Understanding what the people of Afghanistan want and how they want to be led, is a huge challenge that requires an open mind and a willingness to listen and understand. Complex operations have also taught us that problems cannot be solved with military force alone. Success will only come by effectively joining up all the instruments of power from across the sectors. Each of these sectors has a part to play if the strategy is to be effective. The trick is to be able to understand the contribution each sector makes and then to harness and coordinate the various sectors to deliver the strategy effectively. This is a major challenge for strategic leaders, who have often spent most of their careers in only one sector.

One of the corollaries of the complexity of the strategic level is the uniqueness of the problems leaders face. Because the strategic leader must deal with hugely complex organisations that are influenced by an enormous array of factors, no two situations are likely to be the same. We are familiar with the concept of the chaotic system through analogy of the beating of a butterfly's wings in South America causing a hurricane in Asia. Like the weather, the real world in which the strategic leader operates is chaotic. Small changes in variables can lead to vastly different outcomes. This means that the strategic leader faces a set of problems that are unique. There are no actuarial tables to refer to in order to determine the probability of one outcome over another.

The strategic level is also particularly uncertain. It sounds trite, but leaders must lead: they must show the way. The future, shaping it, understanding it and changing it are therefore intrinsic parts of leadership. But the future is never certain, and the further we look into the future the harder it gets. Strategic leaders have the hardest job of all. Strategic problems are long term and the decisions taken have far-reaching consequences. All leaders face the challenge of an uncertain future, but strategic leaders face the greater challenge of having to look much further into the future.

CHALLENGES OF THE MODERN WORLD

Having established that the strategic level is characterised by complexity, uniqueness, uncertainty and ambiguity, it is important to consider the effect of the modern world on the strategic leader. While commentators might argue about the extent of globalisation, the world of the twentieth century is nonetheless more interconnected and more interdependent than at any other time in its history. The recent international financial crisis has shown how small ripples in incredibly complex systems can spread rapidly and have an effect right across the world.

The interconnected and interdependent nature of the modern world gives rise to strategic problems that are also interconnected: global warming; international terrorism; human migration; and the global economy are all problems that cannot be solved by any single country. Addressing these problems requires interconnected solutions; multinational and multi-sectoral approaches are the only viable means of tackling the greatest strategic challenges of today and of the future.

The revolution in communication technology in particular is a double-edged sword for the strategic leader. On the one hand, it has made it easier for a strategic leader to access information and exercise closer control over people and activities far away; and it has made it easier to bring people from different sectors and countries together to develop the multinational and multi-sectoral necessary to solve today's strategic problems. But, on the other hand, the rapid flow of information that the communications revolution has enabled, has also made strategic leader's task more difficult and presented significant risks and challenges.

Strategic leaders are bombarded with information. Some of it is filtered through bureaucracies, with their inherent biases, but lots of it is not filtered at all. This unfiltered 'stuff' can be particularly dangerous for the strategic leader as it is often eye catching, sensational and distracting. The quantity of information available to strategic leaders also presents them with a serious challenge. There is far more information available than can be processed or understood by any human, and in a few seconds it has changed. Understanding what is important and what can be ignored is a greater challenge than the problem of an absence of information that strategic leaders have faced historically. The mesmeric qualities of grainy pictures of newsworthy events taken with mobile phones and uploaded within minutes to a worldwide audience can utterly overload the strategic leader; and the abundance of real-time information, often without context, is a significant challenge for them. The risk of information overload leading to paralysis is very real. When humans are faced with an excess of information, they react by using mental shortcuts to make the problem manageable. The risk is that these shortcuts introduce biases or result in strategic leaders giving greater weight to information that fits their view of the world or what they expect to see. Countering these natural human tendencies is an ever-present challenge for the strategic leader in the modern age.

The revolution in communications technology has also enabled a phenomenal growth in media outlets, and has given a significant voice to groups that previously had no routine or effective means of being heard. Alongside this is the development of a highly competitive news media sector with an irrepressible appetite for instant and sensational news. The reality is, however, that the number of newsworthy events is limited, and the 'gaps' are filled with comment of varying degrees of quality and utility. Furthermore, the news cycle now is so short that strategic leaders are often presented with the 'facts' from news organisations long before their own organisations and systems have reported the incident or developed a line to take. For a strategic leader wrestling with the complex, uncertain, ambiguous and long-term nature of his environment, the risk of being knocked off course or distracted is extremely high.

TACTICAL LEADERSHIP ON STEROIDS?

Some will argue that while they accept there are particular challenges associated with the strategic level, strategic leadership is just the same as leadership at other levels – it is just a bit more difficult. We believe that leadership at the strategic level is different and requires skills that are not necessarily apparent or tested at lower levels. At lower levels, leaders tend to face 'closed' or bounded problems. The environment in which they are asked to lead is controlled, to a large extent, by the organisation in which they operate. The strategic leader, on the other hand, must work in an environment that has a potentially infinite number of variables and where the problems are often unbounded, open-ended with few, if any, precedents.

Strategic leaders are working on the boundary between the organisation they are responsible for and the broader environment. To lead on this boundary, the strategic leader must be able to see and sufficiently understand the whole; and we know this is complex, uncertain and ambiguous. It is inhabiting and working at this boundary, makes strategic leadership so different. It is not just that the strategic leader must understand the environment and see through the complexity and ambiguity, but they must also understand that their actions can, indeed should, shape the strategic environment as well as the organisation they control. As a result, strategic leaders face what some commentators describe as 'wicked' problems. These are unique problems where the very act of taking action can change the nature of the problem.

One of the key roles of a strategic leader is to understand how to influence the environment to create opportunities. A strategic leader must steer the path between the opportunities that the environment presents and their organisation's capabilities. They must shape the environment so that their organisation can exploit the opportunities, but they must also shape their organisation to make the most of the opportunities that are there or which they are able to create. The strategic leader has,

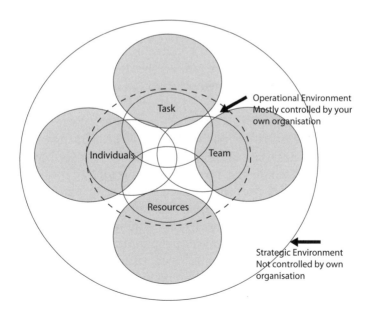

Figure 4.1   Strategic Leadership

then, to be able to think from the outside in and from the inside out. They must also have the moral courage to make their decision and execute their strategy.

NETWORKS, BRIDGES AND PEOPLE

Influencing and exploiting the environment requires a strategic leader to have a good network. With this network they know what is going on and can develop a 'feel' for the environment, where the opportunities might lie and what influence they might have. Building the network requires them to build bridges across a wide array of organisations. With a good network the strategic leader can understand alternative perspectives and has a means of exercising influence over the environment to create opportunities. We have already noted that many of the problems a strategic leader faces are multi-sectoral and often multinational. A network that spans sectors, nations and cultures allows a better understanding of these alternative perspectives and allows the development of more effective and interconnected solutions to such problems.

Leadership at any level is always about people. At the strategic level, a leader must be able to influence those in the broader environment as well as leading the people they are responsible for. The first challenge for a strategic leader is the size of the organisation they must lead. Motivating and aligning large numbers of people, often spread across the world, is particularly difficult. The second challenge the strategic leader faces is the need to lead people into the unknown. At the strategic level a leader is faced with unique and open-ended problems. The path they choose for the organisation will be new and will be based on their judgement about the best course of action. Motivating and leading people along this path requires the strategic leader to be able to inspire people through their vision. Finally, a strategic leader needs to have the skills necessary to influence and shape people across national, cultural and sector boundaries.

RISK AND TIMING

We know that the strategic level is characterised by uncertainty and ambiguity. We also know that the stakes are high. Risk is therefore inherent in the environment and decision making of a strategic leader. The difficulty for them is the uniqueness of the problems and situations they must face. And precisely because these problems are unique, there is no statistical evidence to support a strategic leader's assessment of risk. The strategic leader has to rely on their intuitive understanding of the risks and opportunities they face. Similarly, the timing of their actions to avoid risk or to exploit fleeting opportunities is critical to success and relies on their intuition, much more than analytical tools.

CONCLUSION

The nature of the truly strategic level presents leaders with significant challenges. The strategic level is characterised by complexity, uniqueness, uncertainty and ambiguity. At the highest level in multinational organisations or governments leaders face problems that are multinational and multicultural, which span traditional sectoral divides: interconnected problems require interconnected solutions, at the heart of which must sit effective strategic leadership.

(5) The Characteristics of Effective Top Strategic Leaders

Colonel R. Rider
Ms D. Di Carlo
Major General P. Nazim
Colonel X. Tshofela

INTRODUCTION

Many readers will already have their own list of the characteristics required to be a successful top strategic leader. We seek to identify and discuss what we hold to be those key characteristics and attributes from a multinational, multicultural and multi-sector perspective. Our approach is to consider particular examples of contemporary successful strategic leaders; examine the specific challenges they have faced and their response to them; and draw out the characteristics and attributes, which have been central to their success. We have selected one figure from the world of politics and two from the military sphere by way of complementing, and in many ways reinforcing, points discussed by Ian

Davis and Lütz Bertling, as leading lights from the world of commerce, in their earlier chapters. We turn first to the world of politics.

NELSON MANDELA

Africa is a continent of challenge and opportunity – especially in the context of strategic leadership. In our view there is no better example of an individual who squared up to the challenge and seized the opportunity to demonstrate strategic leadership on that continent, but whose impact has been truly global, than of Nelson Mandela.

At the heart of strategic leadership today in Sub-Saharan Africa in particular is the quest to break out of the cycle of poverty, disease and lack of adequate education, which has beset the region for many decades, so as both to improve the lot of individual states and to help establish the region, as part of the wider continent, as a stronger and more influential entity within the international order. The firm belief within the region is that these problems are not intractable; they can be surmounted by strategic leaders with enlightened vision and resolution. A recurring theme of the more recent past has been the inability of many regional leaders to make decisions untainted and distorted by their own sheer self-interest, with the resultant imposition of inefficient systems, procedures and practices that have failed to provide adequate security and prosperity across their populations. But today, we recognise an increasing desire across the sub-continent for leaders who epitomise the following characteristics:

- Decision-making ability and integrity that transcend personal self-interest and put national (and by extension regional) strategic interests first.

- A high degree of moral and ethical integrity and fibre: the wherewithal to do the 'right thing'.

- A social consciousness that seeks to create harmony between men and between men and nature.

So with that as our backcloth and context, let us consider Nelson Mandela specifically. Mandela, as the first president of the new democratic Republic of South Africa, made a decision of global resonance when he resolved to sweep aside the inequalities suffered under the apartheid regime and replace them with a striving for the creation of a non-racial society, which would respect the values of individual liberty and equality of opportunity and transcend personal self-interest. He faced enormous issues after his release from prison in February 1990: how to prevent South Africa degenerating into a disastrous civil war between blacks and whites; how to work towards the peaceful establishment of a non-racial democracy; and how to consolidate his own position within the African National Congress – a crucial political platform for him.

One of Mandela's greatest attributes was the capacity to develop a vision of a nation that would emerge to value the contribution of each of its citizens both to it and to humanity as a whole. Accordingly, he set himself and his nation the goal of contributing to the promotion of peace and security, and to the rejection of violence and intolerance in their various forms. Mandela's strategic vision for South Africa underpinned a long-term plan of fundamental change for his nation. This called for an acute sense of time and timing, reinforced by strategic patience plus resolution and resilience: all of these were critical to the successful implementation his vision.

Mandela has set something of a moral standard against which other African, indeed world leaders might be judged. While his earlier life has been the subject of criticism in some recent biographies of him – and he himself admits in his autobiography, 'Long Walk to Freedom' that he 'made missteps along the way' – he built up an enormous stock of moral, as well as of political authority both domestically and internationally during his incarceration and in his attitude and actions following his release from it.

He had originally based the operation of the ANC's Armed Wing (the MK) on the principle of non-violence inspired by Gandhi's of satyagraha (non-violent resistance). His clear vision of the just society he wished to see was outlined by his closing statement at the Rivonia Trial in 1961: 'During my lifetime I have dedicated myself to the struggle of the African people. I have fought against white domination, and I have fought against black domination. I have cherished the ideal of a democratic and free society in which all persons live together in harmony and with equal opportunities. It is an ideal which I hope to live for and to achieve. But if needs be, it is an ideal for which I am prepared to die.' Mandela went on to argue that the MK would sabotage symbols of the apartheid regime, while at the same time seeking to avoid casualties in the pursuit of such action.

He did not waver in his conviction or articulation of his vision throughout his time in prison, as exemplified by his refusal of the conditional release offered by the South African President, PW Botha, in 1985 if he were to renounce violence: 'What freedom am I being offered while the organisation of the people remains banned? Only free men can negotiate. A prisoner cannot enter into contracts.' His consistent and resolute resistance to the apartheid afforded him, then, significant political authority; his willingness to sacrifice his own freedom and to endure many hardships over many years in order to achieve a higher goal: liberty and opportunity for all in South Africa, afforded him moral authority. These together were the twin pillars of his success.

His vision of a non-violent South Africa, in which all races enjoyed equal respect and lived in harmony and tolerance was put sharply to the test in 1993, when the assassination of the ANC leader, Chris Hani, could have led to a potentially catastrophic outbreak of inter-racial violence. Mandela demonstrated forthright leadership and his political and moral resolution in his appeal to the population of South Africa: 'Tonight I am reaching out to every single South African, black and white, from the very depths of my being. A white man, full of prejudice and hate, came to our country and committed a deed so foul that our whole nation now teeters on the brink of disaster. A white woman, of Afrikaner origin, risked her life so that we may know, and bring to justice, this assassin... Now is the time for all South Africans to stand together against those who, from any quarter, wish to destroy what Chris Hani gave his life for – the freedom of all of us.'

Mandela's strategic sense and vision, alongside his sharp intelligence and his political and moral authority were evident in his negotiations with FW de Klerk for an interim constitution to provide for the first-ever non-racial election in April 1994, which the ANC, led by Mandela, won with 62.5 per cent of the vote. Furthermore, he demonstrated his keen sense of timing and pragmatism when he formed the initial coalition government of national unity, which included

FW de Klerk as Deputy President to enable the radical, yet peaceful transition in South African political life. De Klerk himself highlighted at the Royal College of Defence Studies in May 2010, the fact that Mandela followed steadfastly the twin paths of reconciliation and reconstruction, and that these were the key elements in overcoming extensive and volatile divisions in South African society, together with the establishment of the Truth and Reconciliation Committee (TRC), whose chairman, Archbishop Desmond Tutu, played a pivotal role in bridging these divides. He went on to praise Mandela for the ability to see and understand the bigger picture and to translate this into a strategic vision – at the heart of which was the imperative to do what he judged to be the 'right thing' – as well as for the ability to seize a strategic opportunity.

Equally striking and resonant is the fact that Mandela had the moral and physical courage to accept the highest level of personal risk and deprivation in the pursuit of his vision: for him the strategic prize of its realisation was of ultimate and overriding merit. Such resolution and tenacity are traits we see in all successful strategic leaders. They also served Mandela well as a platform to build and maintain a constituency of support crucial to his success. Moreover, he was able to organise his legacy and manage the timing and process of his stepping down from office as state president to effect a smooth and effective transition of power in order to sustain the pursuit of his strategic goals in the interests of South Africa and its peoples as a whole.

And so to say that Mandela – the man and his vision – has had enormous impact across South Africa would be undeniably true; but equally, it would be only half the story. What is clear is that the force of his strategic acumen transcends the purely national. Mandela indeed set the goal of achieving fundamental political and social change in South Africa, but there was much more to it than that: he recognised a role for his country to act as a champion of the principle of a multiracial, ethnically harmonious and tolerant society that went far beyond the reconciliation and integration of black and white – a role, then, for South Africa as a lighthouse of global illumination and enlightenment.

That said, Mandela knew only too well that the realisation of a strategic vision – and making it stick – places huge demands and expectations on the strategic leader. When the deaths of Chris Hani and that of Eugene Terreblanche seventeen years later, posed a potentially disastrous threat to the well-being of South Africa, he demonstrated unflinching resolution and the preparedness to give prompt and unequivocal leadership, throwing his personal authority and credibility behind his call for calm and moderation: a call that won the day. Had it not, there would have been the stark prospect of South Africa plunging into chaos with Mandela's vision and hard-won achievements turning swiftly to ash – and casting its cloud across the continent and beyond.

We now turn to successful strategic leaders from the military sphere – one from the UK and one from the USA.

GENERAL SIR DAVID RICHARDS, BRITISH ARMY

General Sir David Richards first came to public prominence as the military face of two intervention operations in East Timor and in Sierra Leone at the start of Tony Blair's first administration, and its so-called 'wars of intervention';

subsequently, and even more prominently, in his role as Commander ISAF in Afghanistan 2006–2007; then as the Chief of the General Staff, and most recently in his appointment as Chief of the Defence Staff in 2010. His time as a senior operational commander witnessed the compression of the traditional levels of military operations across the board: from the strategic to the operational and tactical level. This has meant, in effect, that the operational commander in the field now has to be able to execute an 'in-theatre strategy', whose coordination, in a multinational context, for example, necessarily has to be undertaken within the framework of the strategic aims of Coalition members and other key international partners as a whole; thereby requiring significant engagement with a range of interested parties in national capitals.

CREDENTIALS AND APPROACH AS A STRATEGIC LEADER

Richards' credentials as a strategic leader were established by the dynamic, confident and intuitive approach he adopted in Sierra Leone and later in Afghanistan. The situation he faced in Sierra Leone in 2000 was uncertain and confused with the Revolutionary United Front (RUF) led by Foday Sankoh seemingly about to topple the government of Sierra Leone under President Ahmad Kabbah, despite the presence of a UN mission – inadequately resourced, it should be said – which had been deployed under the authority of UNSCR 1270. Richards' forces were committed by the UK government for an operation to evacuate British (and some other national) citizens (a 'NEO' operation). However, the deployment of a powerful UK military force introduced a new dynamic into the situation in a country which Richards turned to his advantage, not only for the benefit of his national mission, but also that of the government of Sierra Leone and of the beleaguered UN mission there. Through careful and detailed coordination with the Sierra Leone government and the UN force, he managed to stabilise the security situation. Moreover, as the result of a series of successful armed clashes with the RUF, he not only proved the superiority of British arms, he also demonstrated to the RUF the inevitability of their defeat.

This success, likened by Professor Gwyn Prins to the 'strategic raids' of Sir Garnet Wolseley in the nineteenth century, is credited with having saved the country from total collapse and attendant wide-scale bloodshed, and having ushered in a sustained period of political stability.

Richards' approach to his missions has been consistently confident and creative, but – and this is critical – also practical. Operating in a situation of the highest uncertainty and facing challenges of direction and planning, Richards fashioned a clear vision of what needed to be done: what he himself calls 'the golden scenario' which provided for his assistance to the UN and Sierra Leone government to coordinate their military, civilian and economic capabilities in order to turn the situation around. He saw a strategic opportunity, and by his own admission, at some risk to his career, he 'went for it'. It paid off – despite the consternation his action had caused in some quarters in London on account of the perception that he was operating beyond his mission, and hence indulging in so-called mission creep. But he had secured the trump card of the endorsement of the then Chief of the Defence Staff, General Sir Charles Guthrie and Prime Minister Blair.

Sierra Leone was a seminal event in the development of Richards' view of strategic leadership. It established in him a firm two-fold belief in the value of possessing a sharply focused strategic vision and the requirement to deploy commanders into theatre who can be trusted always to act upon their critical judgement – and crucially to empower them to do so. Any commander without such trust should be removed: there is simply no room for compromise here. Richards is crystal clear: in an increasingly interconnected world, personalities matter; individuals with strategic vision and the ability to inspire its achievement can have global impact.

Richards' extensive experience has, then, pointed up forcefully the power of personality, alongside which sits the fundamental and complementary importance of effective information operations and communication of key messages across the spectrum of today's media – and of course this includes making full use of the appeal of the individual where this stands to add force and impact. Richards is also at pains to underscore the need for detailed lateral thinking together with reversing the telescope to see matters from the perspective of the adversary, so as better to devise an approach to secure success.

When it came to Afghanistan 2006–2007, Richards faced an even more complex situation than he had in Sierra Leone several years earlier. His core mission was to help achieve sufficient stability and security to act as the keystone of a sustainable system of good governance. A significant element of this mission was the effective coordination of a plethora of international organisations and agencies whose activities and agenda in practice were not always without a competing dimension. His task was made all the more demanding and involved given the challenges confronting the presidency of Hamid Karzai; not least of these were the resurgent forces of the Taliban.

Richards' success in pulling together the original US mission from 2001, 'Enduring Freedom', the NATO ISAF mission and the efforts of the wider international community, whereby he established coherence and focus in the mission and earned great praise from the then British Ambassador in Kabul, Sir Sherard Cowper-Coles, who noted: 'David Richards' greatest strengths as COMISAF were that he exuded confidence, and that he communicated that confidence to the Afghans and to the international community in Kabul. The commanding general of a great international coalition in war is, whether he likes it or not, fulfilling a highly political role. A counter-insurgency campaign is even more political than conventional war. Success means therefore using many of the leadership skills of a politician, in the best sense of the term: not for partisan benefit, but in order to show a large and diverse Afghan and international audience – many of whom have legitimate concerns about the nature of the mission – that success is not only possible, but probable.'

At the heart of the insights Richards has derived from his operational experiences are those of understanding the type of conflict one is engaged in and then resourcing the response to it adequately. This requires deep and honest analysis: facing up to facts and issues – not ducking or distorting them and a flexibility of both mind and approach; and with all of this unified by alignment with the strategic vision.

For Richards, the principal personal qualities and attributes required of the modern strategic leader coalesce around:

- **Moral courage:** the ability and readiness to trust one's intuition – an intuition borne of personal experience and intellectual reflection, and to apply resolutely one's sense of what is the 'right thing' to do – taking due risk as necessary and whenever to seize the strategic opportunity.

- **Championing of team work:** at the centre of this are the consistent demonstration of a selfless respect for others (Richards bans the 'I' usage) and the valuing of emotional intelligence; all by way of facilitating both motivation and empowerment.

- **Cool handling of pressure and enjoyment of one's work:** these are of deep significance; they help set the right tone and rhythm for the operation of the team/workforce through the creation of a sense of value and purpose. As such, they act as hugely powerful force-multipliers available to those whose responsibility it is to lead.

GENERAL DAVID PETRAEUS, UNITED STATES ARMY

Our final example of a successful top strategic leader is also drawn from the military environment, but in this instance from the US Army: General David Petraeus. He is extensively recognised across military and political circles as an incisive and innovative thinker: he led the team which wrote the US Army's approach to counter-insurgency, (Army Field Manual 3:24). He then proved his abilities as a strategic leader at the cutting edge of battle when he put this into practice while leading the 'Surge' in Iraq 2007–2008. This is widely credited as being a contribution of major significance to efforts for the stabilisation of the political and security situation in Iraq. This in turn crucially underpinned subsequent strategic planning for US and Coalition troops, including both the drawdown of these forces and the key political objectives of building up Iraqi political and governmental institutions and security forces.

General Petraeus inherited a complex political and security situation in Iraq when he assumed command of Multinational Force Iraq in 2007. Having served twice before in Iraq, he recognised that the nub of the issue was the achievement of a political solution to what was, in essence, a political problem: a solution that would rest on the twin pillars of security and stability.

He readily appreciated the necessity to form a series of key relationships. First, with the US Ambassador to Baghdad, Ryan Crocker, to provide for unity of civil/military effort – a 'comprehensive approach' both to problem-solving, command and control of the key organisations in theatre. As Petraeus himself stated at RUSI on 9 June 2010, 'There are no purely military solutions, the military can only contribute to a political solution, the main effort was placed on the political line of operation, so it follows that there is a requirement for a joint approach.'

He also nurtured close cooperation with the Iraqi Prime Minister, Nuri-Al Maliki, reflecting an 'Iraqi First' approach, which recognised Iraqi legitimacy and sovereignty, and the attendant imperative for the Iraqis to set-up their own self-sustaining, representative political and security institutions and organisations. One of the most successful aspects of this 'Iraqi First' approach was the reconciliation and reintegration efforts in Anbar Province, which saw the inclusion of Sunni tribes and former opponents of the government into the security structures of Iraq. This served to undermine the support for fighters of the former regime and for Al-Qaeda who had played on the Sunni-Shi'ite sectarian divide.

Petraeus' adept and astute use of communications and communications technology was clearly evident in Iraq: he ensured that there was a tight connection both to the US government and to public opinion through, on the one hand, weekly VTCs with the Bush administration and, on the other, a prominent information operations campaign to convey his core message that the Iraqi government and security forces with the assistance of Coalition forces would win through and defeat the extremists, not least Al-Qaeda in Iraq, and ultimately achieve the necessary stability and security for the state to function effectively.

His implementation throughout US forces in Iraq of the counter-insurgency doctrine, whose formulation, as we have noted, he spearheaded, led to what was described as 'transformation in contact' of the US Army. This generated a much more flexible and nuanced approach to operations across the Iraqi operational theatre. That the US Army was able to transform itself from a purely 'war-fighting' entity to an effective counter-insurgency force, while remaining in-theatre, was no mean feat by any measurement and stands as testament to the clarity of his vision and his leadership skills in effecting crucial change.

Drawing on the discussion of the preceding paragraphs, on our personal experience of having worked under the general's command and having taken views from officers who have worked closely with him, including in Baghdad in 2007, the following personal characteristics and qualities of the man as a strategic leader emerge starkly.

VISION AND INNOVATION

He articulates his vision in an unambiguous manner. He did so in Iraq despite significant scepticism in certain political and media circles regarding US operations there, and the immense complexity of the situation on the ground. Against this backcloth he set out the desired strategic ends, but was flexible in applying the ways and means to achieve them: for example, brokering deals and concessions with the Sunni tribes in Anbar Province and dealing in an agile and adroit manner with the Shia militias in Sadr City, to achieve in the short term workable political solutions, which were sustainable in the long term.

Petraeus' innovation is demonstrated by his consistent adoption of a 'population-centric' approach, exemplified by his vigorous championing of financial investment in the Kurdish areas of Iraq as a means of assisting the establishment of good governance there – indeed, 'money as a weapon' is something of a mantra for him. Furthermore, this approach pointed up his sharp recognition and appreciation of a change in modern warfare from 'nation-on-nation force' to 'war amongst the people'.

INTELLECT, KNOWLEDGE AND UNDERSTANDING

He couples a first-class intellect with a keen appetite and capacity for deep and wide knowledge of relevance to his role and responsibilities. But more than this, he then channels these to develop a critical understanding of the environment and its issues and challenges in which he operates, so as better to formulate and apply an effective response to them.

TEAMWORK

Petraeus' ability to build and sustain formidable teams is founded on his inclusive approach to tasks and problem-solving; his motto is 'none of us is smarter than all of us together'. Unsurprisingly then, he is an accomplished listener: he takes views from privates to presidents, Shia to Sunni, Kurd to Arab. The coupling of a readiness to listen with well-developed emotional intelligence engenders a sense of personal value and strong esprit de corps on the part of those who work with and for him.

COMMUNICATION(S)

Petraeus' testimony to Congress in September 2007 on the progress of operations in Iraq, revealed top-drawer oral and written communication skills able to withstand the pressure of severe scrutiny. By way of complement, he also understands the vital role in conveying his strategic messages to target audiences in which information operations play in a world increasingly driven by information, communications and the media; one which is technologically advanced and high paced. And on a routine, practical level within his commands, he ensures his strategy is understood and implemented by harnessing the use of the latest technology with the traditional technique of direct, personal engagement – 'walking the talk' with commanders and troops alike. In doing so he creates a culture which, far from undermining or defeating strategy, dynamically delivers it.

ENERGY

He exudes great energy. The former Chairman of the Joint Chiefs, General Henry Skelton, put it aptly when describing Petraeus as 'a high-energy individual who likes to lead from the front, in any field he is going into'. It is this energy, coupled not least with his commitment, determination and strength that contributed to the praise Secretary of State for Defence, Robert Gates, offered to him in September 2008: 'History will regard you as one of your nation's greatest battle-captains.' Furthermore, it underpins an ability that General Sir David Richards admires and lauds, namely one that galvanises individuals and teams across the military and political spectrums – domestically and internationally, whereby they come to share a strategic vision and pull in the same direction for its realisation.

CONCLUSION

The three studies elaborated above serve as their own prism through which one can discern and observe particular characteristics of top strategic leadership. What emerges with striking force from them is the crucial role personality plays in such leadership; the power of the individual as the agent for the fashioning of a strategic vision and of its delivery and sustainment. There is a common thread of clarity and incisiveness of thought and action – all underpinned by pronounced practicality and pragmatism.

These are leaders all about achieving strategic change – and making it stick. The 'sticking' is crucial: it comes from a reaching out across differences and divides to secure a unity of understanding and purpose. It is inclusion, rather than imposition, that delivers the lasting strategic goal.

6 Defining the Context
Top Strategic Leadership in a Multinational and Multilateral Context

Brigadier R. Toomey
Major General K. Engelbreksten
Colonel P. Phelan
Colonel P. Rutherford
Major General S. Storrie

'There is at least one thing worse than fighting with allies – and that is to fight without them.' (Winston Churchill)

'Politics is the art of the possible.' (Otto von Bismarck)

INTRODUCTION

This chapter considers the challenges of strategic leadership in multinational and multilateral ventures, arguing that this context is the norm in international affairs. It describes how participants come together more or less formally into coalitions led by a strategic leadership that finds that it must maintain the coalition concurrently with fulfilling the task. The coalition is a dominant feature of the strategic landscape, shaping the whole approach to strategy and the leader's responsibilities.

THE NORM RATHER THAN THE EXCEPTION

In international affairs it would be rare for a state to gain and keep international legitimacy if it were to conduct a completely unilateral strategy that involved military forces, and was not about immediate self-defence. Strategic endeavours are usually multinational: most states cannot operate alone and even the few that can generally choose not to, as they need to demonstrate the legitimacy of their actions to their electorates and other states. To do this they form coalitions, either from existing alliances (e.g. NATO) or other bodies (e.g. EU, OAU) or on an *ad hoc* basis, often under the UN flag. Moreover, for those endeavours that have any military involvement (or international sanctions), a United Nations Security Council Resolution is considered essential to confer broad legitimacy.[1] Additionally, crises warranting an international response are generally so complex that no organisation is capable of conducting its strategy independently: the solutions must be multidisciplinary. It is almost axiomatic that strategy at this level and in this context is multinational and multilateral; complex with a high degree of ambiguity: hence, in essence, political.

COALITIONS OF THE VARIOUSLY WILLING

The nature of the issue – its challenges and opportunities – influences the type of coalition. Each task and each coalition is unique. Above the individual motivations of participant states

---

1 Although the example of Kosovo shows that this may not always be the case.

and organisations, each coalition's unique nature is shaped by other factors including: the personalities of key actors; the relationships between participating nations; and the cultural differences of its elements. If the issue and its prosecution/ resolution are perceived to be strongly associated with the national interests (that is to say those things that affect the security and prosperity of the state) of one or more states, then those states are likely to commit resources, accept danger to their people and seek to lead or influence the strategy. At the other end of the interest spectrum, when perceived interests are not threatened, perhaps because the issue is largely humanitarian, states are often reluctant to commit significant resources or allow their people to take physical risks. So in Afghanistan and Iraq, the international community's effort has been dominated by the United States, but in the Congo (the UN), Bosnia and the initial intervention in Chad (EU) international organisations rather than coalitions of the willing took the lead at the outset. When danger is high but interests are perceived to be low, the international community is often reluctant to intervene, or holds back, sometimes with disastrous results: for example, in Bosnia in the early 1990s and in Rwanda in 1994).

States engage in multinational ventures for a variety of reasons. Some engage primarily because of a shared commitment to the goals of the strategy, but others may wish to remain in step with other long-term allies; be committed to the success of the international institution; want to increase their standing in that institution or want the international institution to contribute to the upkeep of the part of their armed forces involved. Of course these motives are not mutually exclusive; rather, they help illustrate the complexity at the heart of a coalition.

Ideally, the different national and organisational cultures of coalition members combine to create a powerful degree of creativity and sensitivity to the cultural challenges of the task itself.[2] However, there is also the potential for cultural differences to become fault lines within the coalition, which severely restrict the scope and effectiveness of the strategy. Perhaps one or more participants are so culturally different to the other members that they are barely able to agree the strategy, or may even seek to veto it. Indeed if they were considered too disruptive they may be excluded from the coalition in the first place. Even so, this sort of 'cherry picking' is often impossible, or undesirable for other reasons. For example, important, influential and effective humanitarian organisations are likely to be present at the task whether the coalition wishes it or not. And certain states may have to be involved whose presence might, on the face of it, endanger the success of the operation, for example, Russia in the NATO force in Kosovo from 1999–2003.

It is not only multinationality that brings cultural challenges. The cultural differences between organisations, for example, military and humanitarian ones – indeed many civil organisations – can result in fundamental differences over strategic approaches. A good illustration of this is Afghanistan, where many humanitarian organisations are opposed to NATO's continued use of military and government agencies to deliver aid as part of a counter-insurgency approach.

A particularly problematic area is that of bodies that are not controlled by the coalition or its members working on or influencing the task. The ability to understand such environments and to operate effectively within them is core

---

2   This was an aspect of the coalition to liberate Kuwait in 1991 led by the USA, with widespread participation by Muslim states.

business for a strategic leader. These bodies can include non-governmental organisations, such as international agencies and charities, private military and security companies and agencies of states operating outside the coalition (from both member and non-member states) – and at times this can even encompass organised crime. Of course the government and authorities of the host state, or representatives of the people, are extremely important: they may not be part of the coalition, but it is unlikely that any strategy could succeed without their consent.

All these factors shape the degree of consensus achievable in the coalition and therefore circumscribe the approach it is free to take. It is rare to have complete consensus except when there is a significant, shared and immediate threat. It would be attractive to build a coalition of like-minded states and organisations, but this could leave out members whose presence is critical to the strategy's success. The trick that the strategic leader has to pull off is to maintain the informal and formal alliance at the same time as achieving the strategy. This is essentially a political task, whether or not it falls to a politician.

WHAT SORT OF STRATEGY?

In any event, strategic leaders face constraints that are often not apparent to those involved in, or observing[3] the execution of the strategy or elements of it. The multinational, multilateral natures of coalitions exacerbate the 'ordinary' strategic challenges of balancing ends, ways and means. Coalition members effectively decide what the means will be by their economic, diplomatic and military contributions. They may

---

3   For example, the media, sending-states' politicians and others in the 'commentariat'.

impose caveats on how their resources are employed,[4] further constraining the strategic leadership in terms of the means available. Ultimately this will almost certainly constrain the nature of the achievable ends.[5] Such constraints are not necessarily a bad thing. Sometimes, in order to limit the goals or achievements of a body (the UN, EU or NATO, for example) or of another state, coalition members will deliberately seek to limit the objectives of the strategy. More constructively, one of the reasons states place caveats and employ resources in particular ways or places is to ensure that their contribution is acceptable at home and that the state gets due credit for it. Although these caveats may be unwelcome, they have the advantage of keeping the state in the coalition.

THE STRATEGIC LEADER IN COALITION

The strategic leader steps into this political obstacle course to achieve the objective within the constraints imposed by the coalition. He or she has to employ the 'art of the possible'. This can seem impossible: there is a risk that the maintenance of the coalition becomes the objective, rather than the achievement of the goals of the strategy. The strategic leader has to make the coalition a means to an end, not just an end in itself. Before examining the challenges of the strategic leader, it is worth pausing to consider who this person is, or indeed whether they always exist.

In situations where national interests are low, where a number of states engage on a more or less equal basis, there is often no

---

4   Indeed, they often go further and direct contingents to clear certain operational or tactical decisions to national capitals.
5   A classic example is the 1991 Desert Storm coalition where most of the Arab and Muslim states would only remain in the coalition if it limited itself to the liberation of Kuwait.

strategic leader to start with, but the states often subsequently appoint an individual such as, for example, the UN Special Representative of The Secretary General or, for example, the EU High Representative in Bosnia-Herzegovina. This may be someone who is influential in their own right (a 'big hitter') and is given support by the states (resources, a secretariat, etc.); or it may be a relatively unknown individual who is given few tools for the job. In situations of great national interest to a particular state, such as in Iraq and Afghanistan, a lead state, in these cases the United States, will often assume *de facto* leadership (which not all parties may agree with). Ironically this may make the identity of the strategic leader less obvious; moreover, it may mean that there is strategic leadership by elite, which will often mean that there are contrasting views of the strategy and an intra-elite competition for power and influence. In Afghanistan today, for example, who is the single 'strategic leader'? Any powerful individual working in strategic elite could define themselves as a strategic leader, but their influence on the strategy will vary by appointment, over time, and in accordance with how successfully they play the politics.

INFORMATION AND COMMUNICATION

Wherever the leadership lies common challenges arise: the design of the strategy; the application of the coalition to achieve the task; and the maintenance of the coalition. Military strategic and operational theory describes the importance of (among other things) protecting the coalition's 'own centre-of-gravity'. At the strategic level this is often thought to be the will of the participating states to continue to contribute to the coalition, because if the coalition falls apart, the strategy cannot physically be delivered and its legitimacy is eroded. Ultimately, particularly in democracies, this commitment depends on the consent of the electorates of the contributing states. This consent is potentially vulnerable if the costs (in money and casualties) are perceived to be high, the electorate does not agree with the strategy or if eventual success is questionable. The contributing states are responsible for communicating with their electorates, but one of the strategic leader's tasks is to communicate with the leadership of the contributing states so that they continue to support the strategy. In *Swords and Ploughshares: Bringing Peace to the 21st Century*, Paddy Ashdown describes how he went about this as the EU High Representative in Bosnia-Herzegovina. His description suggests that he spent as much of his time maintaining the coalition in capitals as actually delivering the strategy in Bosnia.

Of course this is not easy. Strategies often take time to develop and are not guaranteed to succeed in their original form. The approach and even the goals of the strategy may vary over time. But we live in an era of impatience, of global media, instant communications and an appetite for drama. Usually the strategic leadership's message is conveyed by the media, a media that acknowledges that it often over-simplifies and exaggerates. In turn politicians in democracies frequently communicate in a similar way. Moreover, they tend to have a timetable and horizon governed by their electoral cycle. The strategic leader in a multinational and multilateral context has to deal with the fact that important participants are often under individual pressure to achieve quick, unambiguous results, when this may not be possible. Strategic leaders not only have to achieve sufficient progress, cohesion and consent, and do so in a virtuous way that is acceptable to their electorates, they have to convince a lot of people that these achievements are genuine and worth supporting.

VICARIOUS STRATEGY

Many contributors to coalitions or strategic endeavours have their own strategic objectives, but lack the power to act independently. What scope is there for a contributing state or organisation to achieve or make progress towards its own strategic goals? Different states take different approaches. Some contribute willingly while making few demands on the coalition leadership, and make little or no effort to shape the strategy. They may essentially agree with the strategy or recognise that they are too minor to influence it. Others do seek to shape the strategy. They can establish conditions of involvement in the first place. They can offer an effective leader and provide that person with some of the tools for the job. They can take a stake in the venture (a military or financial contribution, for example) and assert a right to shape the strategy to reflect their equity (usually on a more or less directly proportional basis). But in situations where a powerful state is in the lead because it perceives that its vital interests are at stake, there may be little that even a significant contributor can do to influence strategy. It can be that certain very effective individuals working within the coalition command and control system in the target country are able to influence execution of strategy in a way that amounts to adjustment of the strategy itself.[6]

CONCLUSION

The challenge facing the strategic leader in the multinational and multilateral context is often one of daunting political complexity and ambiguity. But in international affairs this

---

6   The best approach may be simply to place good people in the right jobs in the hope that they can have this effect.

is the norm rather than the exception. The coalition is the international tool of choice for developing and delivering strategy, but it is not a particularly precise or effective one. Rather than 'only' having to tackle the strategic task, in this context strategic leaders have to deliver and maintain the coalition as the first, critical and enduring task of their strategy.

⑦ Decision Making in the Strategic Environment

Group Captain R. Knighton RAF
HRH Prince Sultan bin Khalid al-Saud
Major General N. Khokhar
Colonel X. Tshofela

INTRODUCTION

The ability to make good decisions is an essential part of leadership. Identifying when the time is right to cease the debate and to choose the right option are significant challenges for leaders. For strategic leaders these challenges are magnified by the complexity, ambiguity and uncertainty that characterise the strategic level. Where strategic decisions are concerned, the stakes are high: strategic decisions really can be a matter of life and death of nations and peoples.

'DECISION' – A DEFINITION

A decision may be variously defined as the act of deciding, a conclusion and the act of making up one's mind. The need or opportunity to move from one condition to another presents the requirement for a decision and makes choice a key component in decision-making.

The word 'decision' comes from the Latin to 'cut off'; in effect to cease the debate and make the choice. Although it is tempting to focus on the moment of decision, a person cannot make a decision without first recognising a choice is to be made, understanding the wider context, the options and conducting some analysis, however superficial that analysis might be. Human cognition and thought processes are therefore essential to decision-making. Although the moment of decision might mark the end point, decision-making is a process of perception, recognition, analysis and choice, even for the simplest of tasks. In recognising decision-making as a process, we can see how the outcome might be influenced by a range of factors which could lead strategic leaders to make poor decisions. By understanding these factors strategic leaders can improve their strategic decision-making.

HOW SHOULD WE MAKE DECISIONS?

Classical decision-making theory requires us to choose the best option having considered and weighed all of the variables and the values of all of the potential outcomes. This 'rational' or normative model requires a decision-maker to understand clearly the goals, know all of the options and be able to estimate the probability and value of each potential outcome. Where decision-makers face relatively straightforward choices in games of chance it might be reasonable to expect people to follow the 'rational' model.[1] In practice, behavioural

---

1 To act rationally a decision-maker must choose the option with the highest 'expected value'. In games of chance all of the potential payoffs and their probabilities of occurrence

psychologists' experiments show that even in these simple situations people do not behave in the way the 'rational' model requires. Some of the most influential research has been conducted by two Israeli psychologists, Daniel Kahneman and Amos Tversky. Both served in the Israeli armed forces. Tversky was cited for bravery as a paratroop captain in 1956 and Kahneman was awarded a Nobel Prize for his work. Their research showed that we employ mental shortcuts (heuristics) and have innate biases, which lead us to deviate consistently from the 'rational' model.

WHAT CAN GO WRONG?

Kahneman and Tversky's research has identified a plethora of potential biases that seem to be hardwired into our brains and which affect us even when making decisions in simple, controlled environments. One of the most significant of the effects identified by researchers is our predisposition to be over-optimistic and overconfident in our predictions. To be able to estimate effectively is essential to strategic decisions, but our predisposition to overestimate the value of our experience or expertise and to be overconfident in our estimates can lead the strategic leader to take excessive risk.

Kahneman and Tversky also found that the way a problem is framed affects our decisions. They concluded that when choices are framed in terms of losses, decision-makers show a propensity for risk-taking behaviour. The researchers also concluded that a decision-maker's frame-of-reference has a

---

are known. A gambler can act rationally by using the laws of probability to identify the option offering the highest expected value. The expected value of any outcome is determined by multiplying the value of the payoff by the probability (expressed as a decimal number) of that payoff occurring.

significant impact on his response. They developed 'Prospect Theory' to explain the experimental evidence which showed that when people are in a 'losing' frame-of-reference, they are more prepared to take higher risks to recover their losses. This theory may explain why there are many more bets on outsiders on the last race of the day.

The effect of people's inability to ignore what has gone before has also been shown to manifest itself in other ways. Richard Thaler, an academic economist and co-author of the book *Nudge*,[2] worked with Kahneman and Tversky in studying these phenomena known as the 'endowment effect'. This 'endowment effect' explains both the extra value people place on something they already own and the strong affinity they have for the status quo. They found that the greater the sacrifice made to gain something the stronger the 'endowment effect' will be. For the strategic decision-maker, this bias should be of particular concern. At the strategic level the stakes are high and a strategic leader may have invested heavily in gaining a position and might therefore fight unreasonably to maintain the status quo. In the military context a leader might have lost lives to win a particular piece of ground and might be reluctant to relinquish that gain, even if a 'rational' analysis would suggest that it is the right thing to do.

COPING WITH THE REAL WORLD

Despite the consistency of the findings of behavioural psychologists such as Kahneman and Tversky, their work is sometimes criticised because of the simple nature of the experiments. The real world, particularly for strategic leaders,

---

2   Thaler, R.H. and Sunstein, C.R., *Nudge: Improving Decisions About Health, Wealth and Happiness* (New Haven: Yale UP, 2008).

is far more complex and uncertain than a game of chance played in a laboratory experiment. Even if a strategic leader wanted to act 'rationally', he could not. There are no actuarial tables to consult or failsafe methods to follow to determine the probability of an outcome. The uncertainty and ambiguity make it difficult to know what the possible outcomes might be, and the complexity makes it impossible, because of our limited cognitive capacity, to consider all of the potential options. As Kahneman put it, 'the failure of the "rational" model is not in its logic, but in the human brain it requires. Who could design a brain that could perform the way this model mandates? Every single one of us would have to know and understand everything, completely and at once.'[3]

To cope in the real world we use our emotions, experience and intuition. We know from neurological research that when parts of our brains controlling emotions are damaged we become slow and incompetent decision-makers even though we can retain the capacity for objective analysis. Because strategic decisions are particularly complex, the need for these shortcuts is particularly acute. In addition, the high levels of ambiguity and uncertainty provide greater scope for misperception and inappropriate interpretation of information. Understanding these influences is important for strategic leaders trying to improve the quality of their decisions.

PERSONAL HISTORY, MORALS AND BELIEFS

Other research has tried to determine how we make decisions naturally. There is no unifying theory of decision-making, but the research has identified a number of mechanisms that we

---

3   Kahneman quoted in Bernstein, P., *Against the Gods: The Remarkable Story of Risk* (New York: John Wiley & Sons, 1998, p. 284).

use to make decisions. These 'naturalistic' theories fall into three broad categories. The first is pattern matching, where we apply previous examples of problems we have faced to the current circumstances and then modify a solution we have used before. The second group is the narrative-based models in which we create a story to explain the current situation and project forward likely outcomes. The final group is the ethical models that indicate our decisions are based on our bedrock beliefs.

These models indicate that our mental images and models of how the world works are important influences on our decision-making. This is not surprising. If we did not make some assumption that the future will, to some extent, resemble the past, we would be virtually paralysed by everyday decisions. The problem we face is how to learn the correct lessons. We tend to construct our accounts based on a small number of variables and simple connections. As a result our learning is often overgeneralised and superficial. Our learning is also biased towards our personal experiences and we draw heavily on events that are experienced first-hand, happen early in our careers or are particularly dramatic. Worse still, we have a strong tendency to learn lessons from success and in doing so we overestimate the effect of our own actions and underestimate the importance of variables over which we have no control. The power of the first-hand experience should also act as a warning to strategic leaders tempted literally to 'see for themselves'. If a situation or problem is complex and the visit short, there is every possibility that what is observed will be more vivid than other information and will be given greater, and potentially inappropriate, weight. After seeing an unusually successful antisubmarine exercise, Winston Churchill is said, for example, to have come away with too optimistic a picture of the efficiency of sonar, which coloured his appreciation of warships' immunity to submarine attack.[4]

Applying our knowledge and the experience we have built up is also fraught with danger. The 'Availability Bias' occurs when people judge the likelihood of something happening by how easily they can call similar examples to mind. If an outcome is vividly imaginable, the probability of its occurrence is likely to be overestimated. This effect has enormous importance for public policy, and is one of the reasons for the overemphasis by the media and public on large-scale disasters such as train crashes. Paul Slovic has written extensively on the subject of societal risk-taking and makes the point that vivid accidents are easier to picture in the mind. This helps explain why detailed safety studies are rarely convincing to the public. As Slovic says, 'the very act of telling somebody about a low-probability catastrophe and going through the fault-tree to show why it's improbable, may make the accident more imaginable and more memorable, and thus seemingly more likely'.

The size and consequent complexity and ambiguity involved in strategic decision-making have particular implications for a number of cognitive effects. One of the ways we cope with this complexity and ambiguity is to simplify the environment by seeking cognitive consistency.[5] Our experience tells us that quite often we like friends of friends and we learn to trust the advice and views of particular people. These types of judgement provide shortcuts to help make decisions quickly and provide a useful filter for the vast amount of information available. In this benign form the search for cognitive

---

4   Jervis, R., *Perception and Misperception in International Politics* (Princeton: Princeton University Press, 1976, p. 245).
5   Etzioni, A., *Humble Decision Making* in *Harvard Business Review on Decision Making* (Harvard Business School Press, 2001, p. 49).

consistency is helpful, but the process can lead us to fit any incoming information into our pre-existing set of beliefs. People ignore 'dissonant' information or become sensitised to the information that supports their mental model, and will therefore seek out consonant information that supports their view of the world. As a simple example of how we become sensitised to particular information, many people will be familiar with the sense that, after buying a new car, there are lots of cars like theirs on the road. The number of cars has not changed, it is simply that our awareness of similar types is heightened, so we 'see' more of them.

'GROUPTHINK'

Strategic leaders must make decision that, by their very nature, have far-reaching consequences. All of the information required to make decisions is unlikely to reside in a single place and those charged with making strategic decisions must rely on others to provide the information required. Groups and the dynamics of groups are therefore key factors in strategic decision-making. The most well-known example of the detrimental effects of group decision-making is 'groupthink'. Irving Janis became fascinated by how bright and shrewd men like John F. Kennedy could be taken in by the CIA's 'stupid' plan to invade Cuba in the Bay of Pigs disaster. In his 1972 book entitled *Groupthink*,[6] Janis presents six case studies of major fiascos resulting from poor decisions made during the administrations of six American presidents. He concludes that the results of 'groupthink' are an overemphasis on consensus, which leads to the selective assessment of information and a consequent failure to critically examine assumptions and options. For the strategic leader, the lesson is one of awareness

---

6   Janis, I., *Groupthink* (Boston: Houghton Mifflin, 1972).

of the fundamental principle that the more amiability and esprit de corps that exists among the members of a group, the greater the danger that independent critical thinking will be replaced by 'groupthink'. Equally, there is also a danger that the fear of challenging the 'official' line or approach in effect allows 'groupthink' to take hold or be sustained.

WHAT CAN WE DO ABOUT IT?

The list of ways in which a strategic leader can err is long, particularly in the strategic environment. Failings can be the result of hard-wired cognitive biases; inappropriate application of experience; and biased perception of information or options as a result of our own self-interest, cognitive failings or group effects. The strategic leader needs to know what they can do about it. There are four broad categories of actions we can take to address these effects.

The first category relates to the heuristics and biases that seem to be hard-wired into our thinking. Although they are likely to have a good evolutionary basis, these hard-wired heuristics are particularly dangerous because the effects are often invisible to us. In such cases forewarned is forearmed: knowing that decisions are affected by our own frame-of-reference and the make-up of problems enables us to test and compensate for the effects of this bias. We can apply alternative constructs to problems and seek advice of others with different frames-of-reference. Similarly, we can adjust for the effect of the vividness and salience of information, by actively questioning our assumptions and seeking alternative and more objective analysis of the probability of particular outcomes. When it comes to our overconfidence in estimating, recognising the propensity for overconfidence, particularly in areas of expertise or experience, provides a first line of defence. We should concentrate on understanding where the extremes might lie, and consciously consider and factor into our estimates hard facts and historic information.

The second category of actions addresses the problems that cause our perception and search for information to be biased. Our earlier examination of the 'naturalistic' decision-making theories and analysis of decision-making in organisations illustrated the effect our values, beliefs and view of the world can have on decision-making. We need to be alert to the effects of emotional attachments that might be distorting our view of the value of a position or option. Developing this understanding requires deep introspection to achieve the necessary levels of self-awareness. Consciously and critically examining our attitudes and emotions, and then testing them against the context of the situation can provide insights into potential biases and help strategic leaders compensate for them. The same also applies to those around the strategic leader. Their opinion and advice may be biased by their beliefs, values, needs and desires. To understand the potential biases in the advice they are being given, strategic leaders need to understand the motivations of those around them. This demands high levels of what Daniel Goleman called, in his seminal work on the subject,[7] emotional intelligence. Those with high levels of emotional intelligence are able to divine the motivations and emotions of others. They form relationships easily, are self-aware and able to control their emotional impulses. Strategic leaders with these skills will be much better placed to identify the warning signs that indicate potential flaws in the information and advice he is being given.

---

7 Goleman, D., *Emotional Intelligence* (London: Bloomsbury, 1995).

A key message of this book is that the strategic leader faces problems that are routinely multinational, multicultural and multi-sectoral. To deal effectively with these problems a strategic leader must be able to work across these boundaries to develop interconnected solutions to these interdependent problems. So, while understanding our own biases is important for success, the strategic leader must also understand the biases that other cultural and national backgrounds introduce. One of the key roles of a strategic leader is to influence the environment in which he operates in order to create opportunities. He cannot do this effectively without understanding how the people he is trying to influence think and behave.

The third category of actions addresses the problems of biased perception, by exposing strategic leaders to alternative perspectives and continually challenging basic assumptions. Janis identified the use of a devil's advocate, outside experts and strong argument as key antidotes to 'groupthink'. Further research has shown that methods that increase conflict, such as dialectic inquiry, produce better recommendations and assumptions than consensus groups. As Peter Drucker, referred to by some as the father of modern management, said: 'the first rule of decision-making is that one does not make a decision unless there is disagreement'. Strategic leaders need to create an environment in which their own views and those of the organisation are challenged and alternative perspectives are given a voice – as Dr. Lütz Bertling so persuasively argues in his chapter earlier in this book.

As well as exposure to alternative perspectives, the nature of the debate in making strategic decisions is also critically important in preventing flawed decisions. The strategic leader has a particularly strong role to play in creating an environment conducive to such debate. Implicit in the adversarial approach is the assumption that the best solution will emerge from a test of strength between the positions, but this approach can lead to battles of power and personality. The goal of effective debate is not to persuade the group to adopt a particular view, but instead to determine the best course of action. This is a subtle but important difference. The most effective debates are about facts, ideas and perspectives, not about personalities.

There are a number of practical steps that can be taken to improve the quality of the debate and avoid some of the biases. Strategic leaders should refrain from declaring their view or position too early, as this can constrain debate and suppress alternative perspectives and challenges. In addition to the use of outside experts and a devil's advocate, the use of 'sub-groups' to break up natural coalitions increases the level of critical debate by cutting across the established power-structure. Similarly, the application of the 'logic of illogic' can be very effective; so, too, the use of hypothetical questions as tools to force alternative perspectives to be considered. The transition to leadership at the strategic level requires a change in thinking and decision-making styles: these need to be less directive, more open-minded and more integration-based. A study of over 120,000 executives concluded that the most successful managers become 'even more open and interactive in their leadership styles ... as they progress in their careers'.[8]

The final category of actions addresses the flaws that result from the way we use our experience, knowledge and intuition to deal with the problems of complexity, uncertainty and ambiguity that strategic decisions present. As we have noted,

---

8 Brousseau, K.R., Driver, M.J., Hourihan, G. and Larsson, R., *The Seasoned Executive's Decision Making Style* (Harvard Business Review, Boston: Harvard Business School Press, February 2006, pp. 111–121).

our learning is often overgeneralised, superficial and biased towards dramatic events that occurred to us early in our careers. Focused introspection can help us understand better what experience and knowledge we draw on to support our decision-making. Armed with this knowledge, strategic leaders can consciously and critically assess their application of experience to the particular problem or context facing them. They can also deepen their well of experience by actively seeking to improve their understanding of the world through vicarious experience and the study of history. Bismarck remarked that 'fools learn by experience – wise men learn by other people's experience'. The crucial point here, though, is that one gains most value from the combination of one's own experience with one's knowledge and understanding of the experience of others. It is the harnessing of this combination of experience that marks out the strategic leader.

CONCLUSION

Decision-making is, of course, an essential component of leadership at every level. Furthermore, strategic leaders are just as likely to be affected by the biases and in-built flaws in human decision-making as anyone else. But the strategic leader faces the particular difficulty that the nature of the strategic environment in which they operate amplifies hugely any flaws in their decision-making – potentially to a global level.

The strategic environment presents problems with multinational, multicultural and multi-sectoral dimensions. These interconnected problems demand interconnected solutions. The challenge for the strategic leader is to understand the different facets in play in order both to influence effectively this environment and to employ all the necessary tools of power to solve the problems it presents. Finally, and by way of an unapologetic 'plug', our own experience has underscored the unique value of a year at the Royal College of Defence Studies as a Member of its Strategic Issues Programme to develop an understanding of these facets as a basis for improving our ability to formulate effective solutions to the toughest of strategic problems.

Developing the Talent
Spotting and Nurturing Future Top Strategic Leaders

Surgeon Commodore P. Buxton RN
Dr S. Akulov
Major General K. Engelbreksten
Major General J. Free
Major General P. Nazim

'What has poor Horatio done, who is so weak, that he, above all the rest, should be sent to rough it out at sea? But let him come; and the first time we go into action, a cannonball may knock off his head, and provide for him at once.' (Captain Maurice Suckling)

INTRODUCTION

From this unpromising start developed one of the greatest strategic leaders of his age – Admiral Lord Nelson. It is in this dismissive summary by his uncle, and his first Captain, that we can discern some of the difficulties associated with spotting and developing strategic leaders. For great leaders can arise from unpromising stock and they will have to lead in a future environment for which the specific talents required are unknown. From going aloft with the sailors, to mastering the complexities of astro-navigation, Nelson, like every other midshipman in the Royal Navy, experienced, and had to master, all aspects of shipboard life. This points up starkly how time has wrought fundamental change in one of the critical aspects in the development of strategic leaders. Nowadays, the leader at the strategic level would be at best hard-pressed to have acquired experience and understanding of all parts of the organisation in which they operate. But they do have to be equipped with the wherewithal to allow them to lead the organisation effectively at the highest level despite this lack. The most successful leaders will be those who can both recognise and compensate for their areas of inexperience. This then is the context for our discussion of spotting and nurturing the talent necessary for successful top leadership.

SPOTTING TALENT

The characteristics of a strategic leader have been elaborated in earlier chapters. So is spotting a future leader simply a matter of identifying the person who displays an abundance of these characteristics? Such an approach makes incorrect assumptions. First, it assumes that these characteristics are present from the outset and that they are easy to spot. Second, it assumes that the best leaders will be those who demonstrate these characteristics most fully.

Some of the characteristics, such as the ability to communicate well, may be obvious early on in someone's career. Others, such as the ability to simplify or to have broad horizons, may be much more difficult to spot. This is not necessarily because the person has not developed these characteristics, but because they have not been put in situations where they have been required or able to demonstrate them.

The second assumption is much more invidious. Some individuals have certain characteristics in such abundance that this can blind others to critical deficiencies or flaws. In addition, for every useful trait there is a 'dark side' caused by its excess: 'action-orientated' can be reckless or dictatorial; 'analytical' can be afraid to act; 'integrity' can be rigid; and 'innovative' can be impractical. It really is possible to have too much of a good thing. The individual who has a blend of all the essential characteristics held in balance, without any of them dominating to the point of distortion, and who has no critical deficiencies or flaws, is a potential future top leader.

The most important characteristics of a leader, as found in surveys of managers, are that they should be: honest, 83 per cent; competent, 67 per cent; forward-looking, 62 per cent; and inspiring, 58 per cent.[1] While sound intelligence may be regarded as a reliable indicator of success as a top leader, this intelligence needs to embrace both the intellectual and the emotional facets. Given the complexities of modern organisations and the problems and demands they face and pose, the requirement for intellectual acumen is obvious; for emotional intelligence it is less so. But a leader's emotional intelligence – their awareness of their own and others' feelings; their ability to see things from another's perspective; and their emotional connection to others – is critical to the success of an organisation. A leader who can sense the various moods not only of their subordinates, but also of all of those with whom they have to interact, and react accordingly, will inspire the

---

1 Cited by Professor Furnham in his presentation to RCDS 2009, *The Dark-side of Leadership*.

collective effort through 'esprit de corps' within a group that is necessary to overcome the toughest of challenges.

The basic criteria for selecting a potential strategic leader are therefore that they are bright enough, emotionally competent and conscientious. To this could be added a list of other desirable characteristics: integrity; courage; knowledge; insight; creativity; and a tolerance of ambiguity.

Many individuals may start their leadership career dealing with operational issues and it is this 'tactical' leadership with which they may feel most comfortable. Hence, as they progress through an organisation, many if not most of them, will retain a preference for these operational issues. Time and again, individuals who have had a brilliant and successful career in an organisation at the operational level are promoted to a strategic leadership position where their previous experience as operational leaders proves to be inadequate preparation for these new responsibilities. Some adapt and rise to the new challenges, but many fail. This is because operational leadership focuses and functions on narrow areas of expertise, with short time horizons and in a known external environment. Strategic leadership, on the other hand, has to contend with a far wider spectrum of issues and of weightier import, over much-extended timescales and often in a dynamic and unpredictable environment.

It is not at all certain that a path, with a series of goals to be achieved, to develop strategic leaders can be defined. Looking at the career development of many of the great leaders suggests that they are characterised by faltering progress and many failures. Churchill's advocacy for the disastrous Dardanelles campaign in the First World War led to his short-term political downfall; any suggestion then that he would subsequently become a great strategic leader would have been met with howls of derision from many quarters.

Can this feature be used to spot future leaders? It would be somewhat perverse to suggest that one should look for people who have failed in order to identify potential leaders. It is not so much the act of failing that is important, but their reaction to it. The point at issue is: do they learn from the experience and try a different approach shaped accordingly? This would clearly be a good response. Or do they either press on regardless or become hopelessly dispirited and give up – and to no positive end in each case? Furthermore, occasional failure may be the price that has to be paid by the nascent strategic leader for their propensity to see things differently and not be 'one of the crowd', but rather to push the boundaries and not be hamstrung by conventional wisdom or approach – to take a calculated risk. After all, 'the only person who never made a mistake, never made anything'.

As we will see later, a leader's experiences, both successes and failures, are more critical to their development than any number of formal training programmes.

One of the quintessential differences between tactical and strategic leaders is that the latter are concerned with the chaotic, external environment in which their organisation operates, whereas the former are chiefly concerned with the relative stability of the organisational environment itself. As the Danish physicist, Niels Bohr, is said to have observed: 'Prediction is very difficult, especially about the future.' The only prediction that can be made with much certainty about the environment in which a future strategic leader will be operating is that it will be different from today's environment.

If we cannot predict the environment how can we select people who will be successful leaders in it? The key here is to choose people who are capable across a broad range of fields, and are equipped with a correspondingly wide spectrum of skills – going beyond those relevant just to the tasks of the moment – and with the flexibility and adaptability to marshal them to the demands of a developing and fluctuating environment. Diversity in this regard is a big plus: ideally an organisation would have multiple potential leaders with different mixes of skills. Just as 'selection of the fittest' works in nature so the future environment will point up from this pool of talent those who are most suited to respond effectively to it.

What we can be sure of is that, in today's increasingly interconnected world, strategic leaders of multinational organisations must be able to guide those of different cultures towards a common goal. The cohesion of an alliance becomes as important as attaining any of the goals that the alliance was set up to achieve. The politics of the alliance may also mean that the selected leader may be the one who is politically acceptable to the greatest number of the multinational partners rather than the one most suited to the task in hand.

DEVELOPING TALENT

If an organisation is to maintain its strategic advantage then the development of the next generation of leaders must be a strategic imperative. It naturally follows that today's strategic leaders in an organisation should be deeply committed to the organisation's leadership development programme. This is not simply a moral calculation to put something back into one's society or organisation, but rather a hard-headed response to the need to ensure continuity of top-quality leadership. Does this mean that current leaders should be personally involved in teaching leadership? Some companies see this as an essential task for their current leaders – Intel and GE spring to mind as two examples of this – while others might argue that not all top strategic leaders can convey their skills to their juniors – not all top leaders are top-drawer teachers. It is probably truer to say though that, simply working with a great strategic leader, even if they are a poor teacher, will convey valuable lessons to an aspiring junior. Of even greater importance than this personal involvement is the leader's responsibility to ensure that the development programme and the associated processes to spot potential leaders, such as a formal feedback system, are embedded in the very culture of the organisation.

An interest in, and an understanding of strategy are rare commodities. To foster and encourage such thinking Sir Jock Stirrup, the Chief of the Defence Staff (the UK's most senior military officer), held 'strategic discussion circles' with his subordinates in order to encourage their thinking, and prepare them for leadership at the strategic level. Such involvement of senior leadership is rare since only the most self-confident leader can take responsibility for the development and mentoring of a future strategic leader. This is because organisations are hierarchical and the encouragement of more junior personnel risks upsetting the other, more senior members of the organisation who, although higher in that organisation, may be less suited for strategic leadership positions.

The identification and development of future strategic leaders is, therefore, fraught with difficulties as it risks opening up many unwanted issues within an organisation. But, whatever the cost, this activity is essential for the long-term success of the organisation.

Any organisation that values its long-term success will sustain adequate investment in leadership development and will make it a core activity within and across organisational boundaries so as to ensure both that this is a function which 'belongs' to the whole organisation and also that those personnel identified as having top leadership potential will be able to gain necessary experience across the organisation without undue restriction from structural boundaries or local narrowness of approach. Such wide-ranging experience is crucial to the development of successful leaders. The purpose of developing these competencies in individuals is not somehow to make them better people, although that may be a by-product of the process; rather it is a means to deliver value to the organisation. That is not to say that such value is necessarily immediate; some experiences, and the competencies they produce, may not deliver their true value until the individual has risen to the very pinnacle of their career. In an ideal world, however, both the individual and the organisation would gain both an immediate and a long-term benefit. By working outside their 'home organisation' an individual not only broadens their experience, but also has the opportunity to build their own network of contacts. The building of such a network – both formal and informal – is a key enabler at both the personal and organisational level.

Some companies regularly set a group of their young managers broad, general problems to solve that are of concern to the company. Even if the solutions they propose are of limited value, the individuals have received valuable experience by working outside their field and at a level significantly above their norm; and at the same time, the company has contributed to the development of its future leaders. In many instances, however, even if the solutions are not fully formed or readily implementable, they generate valuable new ideas that can be further developed. The exercise can therefore have both short- and long-term benefits for the company.

While operating within different organisations and sectors will broaden an individual's experience, it may also show that, while the tactical leadership of them may differ greatly, there is more commonality than difference at the strategic leadership level.

There are a large number of leadership training courses, many of which provide excellent teaching of the skills required for tactical leadership. But can strategic leadership be taught? As we contend above, a strategic leader needs to have some basic characteristics and to have acquired an appropriate level of experience and understanding relevant to the organisation they lead. This is not a short-term process. Any notion that a course can substitute for this long-term effort and inherent ability is nonsense. While a course can teach you about strategic leadership and provide some of the techniques that are used, it cannot teach you to be a strategic leader.

Effective leadership is not an instinctive ability, rather it is a skill that can be developed and honed through appropriate training and guidance. While at the tactical operational level such development is within the scope of most individuals, at the strategic level the individual must have some innate qualities; such individuals are much fewer in number. Spotting individuals with such qualities is clearly an important task for any organisation. There is no special set of techniques required to accomplish this: they are the self-same ones that are used to develop talent throughout the organisation. Thus, at all levels in an organisation, there needs to be a systematic feedback process: mentors and coaches must be used both formally and informally to encourage juniors to develop; and the study of leadership must be actively encouraged. These techniques, especially where they involve personal feedback, are not without their difficulties. Giving and receiving personal feedback can be a very difficult process that can result in significant tensions in interpersonal relationships and a reluctance to engage in the process. Difficult or not, it is an essential process and the failure to give honest feedback to a person is often worse than not giving it at all. At the operational level, the context of leadership development may have as its starting point the environment of the organisation in question. However, at the strategic level any such development must necessarily be anchored within a vastly more complex external environment, where multinational, multicultural and multi-sectorial issues must be given equal prominence with those relating to the organisation itself.

As well as receiving formal training, leaders also need time away from the pressures and immediacy of the operational environment in order to have time to think and to reflect on their own development and on the larger, strategic picture as its backcloth.

If there is a plethora of courses on leadership, there is matching number of books on the subject. We have put forward the argument in this chapter that strategic leadership cannot be taught merely by means of training courses; but can it be learned from books? Those books which treat the techniques of leadership by focusing on a single individual run the high risk – irrespective of the extent of that individual's success – of taking too narrow a view. They miss the essential point, namely the imperative to develop one's own approach to leadership. The good leader is one who can synthesise elements from different role models and not just slavishly copy one of them. Hence, while books about *leadership* in the abstract are unhelpful; those books, on the other hand, which illuminate the qualities, experiences and insights of successful top leaders can be an invaluable source of inspiration and guidance. It is in reading about these things that we can learn what formed such leaders and the reasons why they made particular decisions. Even though their precise circumstances will not wholly match our own, a knowledge of a wide range of leaders, and the problems they faced, will provide a sound basis for one's own decision making.

CONCLUSION

Identifying individuals with the potential to become a strategic leader early in their career can be extremely difficult. As the external environment in which the future leader will be working is unknown, an organisation should develop a number of potential leaders so that when the time comes, the most appropriate ones can be selected. Training can help to refine their leadership skills but is no substitute for both experience and expertise relevant to the core business of the organisation and an innate ability for top leadership.

The long-term health of any organisation is dependent on the identification and development of its future strategic leaders. Developing those who will lead tomorrow is a key task for those who lead today.

PART IV

'Golden Thread' and Unifying Themes

Conclusion Top Strategic Leadership: The Critical Difference

Charles Style

'Turning and turning in the widening gyre
The falcon cannot hear the falconer;
Things fall apart; the centre cannot hold;
More anarchy is loosed upon the world,
The blood-dimmed tide is loosed, and everywhere
The ceremony of innocence is drowned;
The best lack all conviction, while the worst
Are full of passionate intensity' (Yeats)

Google 'strategy' and you'll find about 854,000,000 entries on offer. Google 'strategic leadership', and the number is 13,700,000. There's no lack of interest in the subject. The question is why?

Part of the answer lies in its resistance to easy definition. Some think that leadership and thinking at all levels are much the same; others believe there are differences. This concluding chapter sets out to summarise some aspects of these differences, in the belief that the lessons are applicable widely.

HUMAN FACTORS: ATTRIBUTES AND LANGUAGE

On repeated occasions when examining the characteristics of great strategists over recent years with colleagues, a consensus has been evident. The recurring themes included moral values and the courage to swim against the tide, a sense of humanity in its entirety, generosity of spirit, conception and delivery of great visions against the odds, comprehension of context, belief in things that matter, conviction combined with self-belief, an ability to see and grab opportunity, thinking 'beyond the battle of the day' (relating to Nelson Mandela), wisdom, and reaching partnership with past enemies for the long term good.

This is intuitively an unsurprising list. And yet in the area of strategy the language we can sometimes encounter is mechanistic, structural, organisational, antiseptic, linear. Anything but human. The Financial Services Authority's early analysis of the reasons behind the financial crisis included concepts like leverage, maturity transformation, and pro-cyclicality. It also included 'misplaced reliance on sophisticated maths'. An excellent two-day Marie Curie research conference at the Manchester Business School in November 2009 ('Understanding the New World of Financial Risk Management') on these same lessons found much of the agenda steered by participants towards the beguiling false hope of 'even cleverer maths'. Some talked in effect of the 'risk-free risk'; I felt as if I was on Mars. A senior banking colleague and I were almost alone in presenting the demanding human dimension of top leadership on a platform we jointly shared.

All who have occupied senior positions at various levels can list common characteristics of the good leader: professionalism, intelligence, integrity, powers of communication and so on. These all apply to top-end leadership too; but at that level there are forces in play which mean that something more is required.

THE CONTEXT FOR STRATEGIC LEADERSHIP: KEY CHALLENGES

First, a truly strategic issue without an international dimension is more or less unimaginable. Every major contemporary challenge – for example, climate change, pandemics, shortage of natural resources, migration, sea levels, terrorism, organised crime and the obstructions to agile cross-boundary commerce including the freedom of the seas – self-evidently requires global vision if it is to be met effectively. And yet the obstacles are legion; indeed centrifugal forces operating in the opposite direction (for example some elements of the Eurozone stresses and hints of protectionism) are increasingly evident. In such a world, the definition of national interest and strategy is a challenge. The environment in which we all operate now is unimaginably complex, usually ambiguous, extremely fast moving, and the enemy of the long view. Everything is universal; it comes to us through social media, the internet at large and 24-hour news. In both public and private sectors, it is harder to keep secrets. This is not just the direct result of the google phenomenon; it also arises from a substantial shift in behaviour. News feeds update by the second, technology feeds a voracious appetite for newness and stimulation, and thus patience is lost: in things, in entertainment and in leaders themselves. Deep analysis suffers at the hands of the sound-bite. This places top leaders in a windy place; the long view can suffer.

And yet in a world moving so fast, the long view is more, not less important. In 20 years' time humanity is going to be occupying a *very* different place; it is to that place that strategy should be directed. To take but one example, if it is roughly true that the population of the world will have increased by 2.5 billion to over 9 billion by 2050 and that simultaneously the proportion of elderly people in the Chinese population will have increased from between 10–15 per cent now to over 30% in 2050, how should we be thinking about the sustainable future of humanity? Of course primary time horizons depend on where you sit. An international company might look at least 5 years ahead; national governments must and do focus to a more distant horizon, while having to handle the realities of challenges much closer in.

CORE DEMANDS ON STRATEGIC LEADERS

These circumstances demand of strategists and top-end leaders that they look up and out, not exclusively down and in. I hope as a seafarer that I may be allowed a Naval example drawn from my days working up ships from several countries in the mid-90s. Commanding Officers would find themselves handling massive information overload and the press of demands close in and further out. Some would stand back, sometimes physically, to maintain an overview while handling personally only the immediate tactical challenges that absolutely required their attention. Others got drawn to the narrow and the close. I would see some in this latter category physically being sucked into their command displays until all beyond was shut out. Of course this is a very tactical example. However, there is much psychology in it which is applicable at the highest as well as at lower levels. The tougher the challenge, the harder it is to retain the strategic shape of things.

Top-end strategists and their closest supporters find great demands placed on them; their qualities and behaviours need on occasion to include the hard-edged ones of which ruthlessness may be one; Winston Churchill sent many hundreds of British service people to their near-certain death in the Second World War in order to protect the Enigma secret. Toughness, resilience, determination, high stamina and clarity of vision all play their parts. They are widely recognised, understood and valued by leaders and followers alike.

Less readily recognised in some quarters are other needs. Strategic leaders need to be self-aware, digging within themselves to levels and making choices which are difficult to reach when handling the imperatives of the close-in. A straw poll amongst past colleagues has repeatedly posed them the question: 'do you think you are more likely to face personal ethical choices in the next 20 years than you would have been in the last 20, or not'. The answer is always the same: 90 per cent go with 'more likely'. Of course they do; such is the nature of the world we live in. And yet how well do we help prepare people for these personal choices? Do we value education adequately, alongside, but distinct from, training and attuned to the understanding of a very different world?

Again culture is central; it must create room for integrity and originality at both the personal and organisational levels.

Toxic cultures can creep up on you; where they exist, ideas and vision get buried. This, however, is not advocacy for anarchy; once a direction is set, drawing on the best and most diverse of insights and reassessing from time to time, the team must follow it. At this point, if the follower can no longer sing the company song, he should go.

Turn back to those recognisable characteristics by which our great strategic leaders were identified at the beginning of this chapter. None is new; do we do enough to raise consciousness of them? The answer to this lies in setting a good example, education and mentoring – from the youngest age, and then continuously through life. In this the Royal College of Defence Studies and the Defence Academy of the United Kingdom, of which it is a part, play a crucial role.

The strategic leader should understand 'the other side', meaning the enemy, the competition, the working culture of the newly acquired subsidiary, or the negotiating partner for example. There will not necessarily be agreement; but there must be a readiness to listen, to understand, and – in the words of one of the finest military strategists anywhere in the world – 'to be persuaded'. This fact, ever more important in an interconnected world, of course places a premium on diplomacy (again broadly defined) in the building of strategy. How often have organisations misunderstood the constraints or the opportunities arising from cultural differences when operating internationally? Education and the creation of physical and conceptual space for thought and originality arising from the broadest range of sources are essential.

In any organisation where major and urgent change is contemplated, a deciding point will be the willingness to change behaviours and cultures.

STRATEGY: ITS EVOLUTION, EXPRESSION AND EFFECT

Key formative factors in strategic analysis include history, geography, culture and leadership. In addition – and notwithstanding the unnecessity to always agree – good strategies provide threads to the future for all involved; this is one implication of 'a sense of common humanity', although the imperative is practical as well. Both or all sides should have a stake – an 'interest' – in the outcome. If 'victory' (however illusory such an idea may be) is imposed unilaterally on the 'loser' in such a way as to bury his fair aspirations and future hopes, then the accommodation is no accommodation at all. Long remembered sense of injustice will bite back. The Israelis and the Palestinians will never agree on the narrative; neither will the Indians and Pakistanis. Neither did the Protestants and Catholics in Ireland, nor the blacks and whites in South Africa. But in two of these cases a passable form of peace was reached by strategic leaders on both sides taking leaps which until the moment of the great step were unimaginable; and yet the next day there was a new reality.

Strategic leaders do create new realities, at whichever level they are given the opportunity to operate. Crucially, opportunities can pass. In other words timing is of the essence; an action may be maladroit one week, the key to success the next, and disastrous the week after that: 'leap through the window of strategic opportunity whilst you can; because it is open rarely and all too briefly'. This sense of timing is one of the key attributes allowing statesmen to 'deliver great visions against the odds and to see and grab opportunity'.

There are differing ways for good strategy to evolve. One is for top leaders to sub-contract its formulation, to have their staff fashion strategy for them. If they then unequivocally take personal ownership of the strategy, proclaim it, lead and drive it, this can work. Much more effectively – and more or less invariably in the great examples – strategy is the product of the leader's personal inspiration. What cannot work, what represents the antithesis of the meaning of strategy, is leadership which demands a strategic plan of its subordinates without consistent guidelines or the necessary unifying concept, and then rejects or fails fully to own either strategy or plan.

Organisations see this every day in change programmes from which non-leading 'leaders' think they can remain at a safe distance by delegating ownership and inspiration. Capacity is needed in organisations not only to conceive but to deliver change; this requirement can be under-estimated, as busy people grapple with in-trays of routine paperwork on top of herculean change to be delivered invariably against silent internal resistance. Thus 'strategic delivery units' and the like in organisations have a place, with one proviso. Their roles must be correctly understood. Are they meant to inform, to create, to monitor, to lead, to deliver, or to police? If there is no confusion, all well and good. If the opposite is true, then it is instructive to ask why? Inevitably in this case the consequences are negative.

Finally it should be possible to express the essence of good strategies very briefly: one phrase, or a sentence. If strategies must be written in huge weighty documents with dense type and annexes, it might be difficult to distil that essence.

To allow people to march to a strategic drum beat, a strategy should be memorable. Its expression needs to encapsulate a sense of the destination and a sense of the action required.

It must enable everyone making the march to visualise how he or she is to contribute to getting there. In some large organisations, people can be chary of stating what this or that strategy actually 'is'. If asked, they might reply that 'it's a document', 'it's a list', 'it has four sections', 'it's the organisation's approach to x or y', 'it's an attempt to ...', 'it was published last month'. If an unequivocal statement is not on offer, the individual probably doesn't know what the strategy is or is frightened of expressing it.

In such circumstances, it is probable that no unifying theme has been defined; therefore there is in truth no strategy. On the other hand if you are Vosper Thorneycroft – 'get out of ship building'; Western leaders in the Cold War – 'mutual assured destruction', 'flexible response' etc; BAE Systems – 'go for the US market'; Allied leaders in the Second World War – 'Germany First', or President Musharraf – 'go with the West', then you have succinctly met the test. I paraphrase boldly and hope I will be forgiven for doing so. I merely suggest these examples had recognisable strategic form.

THE 'CHALLENGE OF STRATEGY'

Why is good strategy now thought to be so difficult to deliver? Many of the answers lie in earlier observations throughout this book; the matter is the subject of hot debate as it goes to print.

Experience and context play their part. Formative military years in the Cold War are instrumental. This point can be over-played. However in the first frigate I commanded, and all in which I had previously served, the most private safe contained 'sealed orders' for war. It was not necessary to specify which war. It was all pre-planned, and these ships knew their own way to the start positions because we practised it every year. They also knew which way to turn when we got there. How extraordinary that sounds now; but such was the context in which a generation grew up across the armed services and amongst their civilian counterparts.

This is not to suggest that life was simple. Dangerous situations arose, and strategy based on deterrence and containment demanded extraordinary clarity of thought and execution. However the lid was kept firmly closed over this period on cookers which had nonetheless remained pressurised since Second World War; in the last two decades, those lids have been released, posing diverse and complex challenges.

Yet ever since the Second World War many Western countries – with the notable and obvious exception of the USA – have been adjusting their projection of power across the globe downwards. In Britain's case the empire disappeared, and with it ever larger chunks not only of the sort of military capability defined by mass but also of the equivalent civil (including diplomatic) capacity. The stark statistics speak for themselves; they do not mean that Britain and other nations have failed to advance; many have singular strengths and important international linkages to this day. On the military side it is argued that Britain is smaller but better, cleverer, more technological, more agile. Even so, perhaps with mass went routine experience of the military strategic level in fluid situations. Put another way, very few military strategists indeed since 1945 were again to experience strategic leadership in the way Alanbrooke did forming Allied strategy during the Second World War with Roosevelt, Churchill and Marshall. Insofar as leaders engage at or near the true strategic level today, it is mostly in expanding international fora within which unified strategic vision is hard to achieve for obvious reasons.

This is to imply no criticism. The British Armed Services, though continuing to get smaller, are amongst the best in the world, and a source of national pride. It is however to suggest that military experience is nowadays primarily gained at the 'operational' and 'tactical' levels (next steps down from strategic). Thus the very best of broad education is at a premium in order to promote both military and cross-government strategy making on the international stage, along with the habit of thinking big, unconventionally and imaginatively. Doing this in a context which for decades has been marked – almost without reversals – by declining budgets as a proportion of national wealth is particularly challenging as Britain's world role has been adjusted. Care is needed to reward originality, and through leadership to head off internecine behaviours and suspicions where they arise.

The case of that remarkable Second World War foursome – Roosevelt, Churchill, Marshall and Alanbrooke – is again instructive. They spent weeks closeted together refining their combined strategies in tough prolonged meetings. Andrew Roberts' account in *Masters and Commanders* of the dynamics makes quite fascinating reading. The point is that they gave the subject *time* and they did it *themselves*.

On the commercial side, there is another factor in play. Statistics show that most major successful companies have a surprisingly limited life-span. Some buck the trend, continually renew themselves and remain on top over decades. But most don't. There is a natural bell-curve; a likely life-cycle. We all know how extraordinarily difficult it is to bring about major change in organisations, especially those which have become over-comfortable, over-easy with the 'right to success'. Many set out on change programmes; only a minority follow them through.

The lesson is transferable. Much can be gleaned from the recent experience of some small- and medium-scale states which have leveraged fresh thought, agility and energy to a remarkable extent and good effect.

Nationally and internationally we need open minds to big ideas. Great strategic leaders invariably embody greatness by virtue of their imagination and ability to look ahead. The challenges are well known: simplistically put, how – for example – are the 'West' and the 'East' going to handle the friction surrounding shifting power balances and economic relationships? This is rich territory for 'unimaginable leaps'. How, crucially, are they going to learn to understand, listen to and trust each other, given that language, philosophy, values and perceptions are so different? In China, for example, two logical inconsistencies can live happily together and be equally true; 'this is real life'. Western reasoning and education lead clear thinkers to rebel against such an idea. In a complex situation arising between representatives of these two traditions, calling for a steady hand and a cool head, it would be best if the protagonists allowed for such differences. Thus people from the West should be spending time in the East learning to understand how people in that region reason, and vice versa. A mere understanding of language is simply not enough.

The military revere certain strategic visionaries of whom by far the greatest, living about 2500 years ago, was Sun Tzu. Do we in the West pause to remind ourselves from which great civilisation he came? Sun Tzu and a small number of other practitioners/thinkers bring an understanding of strategy which is worthy of attention well beyond the military sphere. Military people attempt to bring order to chaos by asserting mantras: for example 'it is all about matching ends, ways and means'. And we must 'assess – plan – communicate – coordinate – implement – review'. Of course in mechanistic terms these sorts of structured routes to answers are useful, yet they are often not applied. They bear repetition, and are applicable to all undertakings at all levels. Good strategy and good top leadership are tough, demanding practical and pragmatic things to achieve, in which choice often lies between the not-too-bad and the appalling.

Even so, beyond the mechanics, the secret to better strategic leadership is certainly not to try to teach *what* to think, and lies only in part in ideas about *how* to think. What we hear repeatedly from the very best practitioners is that personal qualities, behaviours and creativity are what open top leaders to true strategic brilliance: who they are *inside*. These qualities include wisdom, along with sound instinct, judgement, open mindedness, and a propensity to give time to reaching understanding with and of friend and foe alike: 'a sense of humanity in its entirety'.

If there is one implicit message in this volume, it is that internationalist education – cross-sector, cross-culture – such as has been offered at the Royal College of Defence Studies over many decades and elsewhere in the Defence Academy of the United Kingdom is beyond value for our common futures. Of all the issues and qualities which such education needs to address, I submit the most important is cross-cultural understanding.

Afterword
Global Strategic Leadership: Reflections Across the Private, Political and Military Sectors

The Rt Hon Lord Patten of Barnes CH

There has never been a better time for the publication of this book. It is a response to a global landscape that has never been more complex or volatile; one which makes for an unparalleled level of ambiguity and uncertainty. It offers unique value in its dovetailed treatment of strategy and top strategic leadership across the private, political and military sectors in a concise, practical and fresh way. Moreover, it provides extensive and deep insights into how to *interpret* and *shape* that environment.

This is a book that demonstrates how the multinational, multicultural and multi-sectoral approach can supply the golden thread of effective strategy and strategic leadership;

how simple insights can be refined out of complexity. These insights are those of practitioners: a collection of contributors who have applied extensive personal experience and expertise coupled with concentrated study and analysis – the product of their time as members of the internationally renowned RCDS Programme – to key areas of strategy and top strategic leadership; in doing so they have reached across the toughest political and cultural barriers.

Its messages chime with my own experience as a politician and diplomat dealing with a range of international strategic issues; not least those of being in the thick of securing a lasting settlement for Hong Kong. This was a process that put personal qualities and political leadership skills sternly to the test. All of us involved were aware that its outcome would be of global strategic significance. The stakes were of the highest; happily, the settlement has stuck. And so today, the quickening shift of power and influence from West to East presents a huge challenge to those at the top strategic level whose responsibility it is to manage this shift to the benefit of the international community as a whole.

This book faces up to the issues and demands associated with such global challenges squarely and it illuminates what it takes to respond to them effectively. All current and aspiring top strategic leaders should read it – whatever their field, whatever their experience and expertise. It is a book whose time is ripe. I commend it unreservedly and with enthusiasm.

About the Editors

Vice-Admiral Charles Style was until spring 2012 the Commandant of The Royal College of Defence Studies, London. A Cambridge graduate, his sea commands included five ships culminating with the aircraft-carrier and fleet flagship *HMS Illustrious*, followed by flag command of the UK Maritime Force, and the maritime element of the NATO Response Force. As Deputy Chief of Defence Staff (Commitments) in the Ministry of Defence, he was Director of UK Military Operations, and ran the Defence Crisis Management Organisation. These and other roles have engaged him in leadership, strategy and capability development both nationally and much further afield. He is an Executive-in-Residence at the Manchester Business School, and has established a business providing international strategic advice, consultancy and leadership development.

Nicholas Beale is a strategy consultant and Chairman of Sciteb思特, a firm that provides innovative strategic thinking to top management both in developing their strategies and their boards. He is known internationally for innovative approaches to strategic problems, including fundamental new thinking on systemic risk, and has given seminars at Harvard, Oxford and Tsinghua Universities as well as to central banks and regulators worldwide. His work has been featured in the FT, WSJ, HBR and *Nature*. He is a Fellow of the RSA and the Royal Institution, a Freeman of the City of London and a Liveryman of the Worshipful Company of Information Technologists.

David Ellery is a Foreign Office official with extensive first-hand experience of strategic policy-making and its implementation at the national and international level. He is also Senior Directing Staff Emeritus at the Royal College of Defence Studies (where his recent work on India and China as emerging global powers has been published) and is a Fellow and Trustee in the University of Durham, where, as a member of Van Mildert College, he graduated with a Double Honours degree and a distinction.

Index

If you have found this book useful
you may be interested in other titles
from Gower

**Complex Adaptive Leadership:
Embracing Paradox and Uncertainty**
Nick Obolensky
Hardback: 978-0-566-08932-9
e-book: 978-0-566-08933-6

**Escalation in Decision-Making:
Behavioural Economics in Business**
Helga Drummond and Julia Hodgson
Hardback: 978-1-4094-0236-7
e-book: 978-1-4094-0237-4

**MisLeadership:
Prevalence, Causes and Consequences**
John Rayment and Jonathan Smith
Hardback: 978-0-566-09226-8
e-book: 978-0-566-09227-5

GOWER

Strategic Review:
The Process of Strategy Formulation
in Complex Organisations
Robert F. Grattan
Hardback: 978-1-4094-0728-7
e-book: 978-1-4094-0729-4

Transforming Government and Public Services:
Realising Benefits through Project
Portfolio Management
Stephen Jenner
Hardback: 978-1-4094-0163-6
e-book: 978-1-4094-0307-4

The Evolution of Strategic Foresight:
Navigating Public Policy Making
Tuomo Kuosa
Hardback: 978-1-4094-2986-9
e-book: 978-1-4094-2987-6

Visit **www.gowerpublishing.com** and

- search the entire catalogue of Gower books
  in print
- order titles online at 10% discount
- take advantage of special offers
- sign up for our monthly e-mail update service
- download free sample chapters from all
  recent titles
- download or order our catalogue